Book 1
Circuit Engineering
By Solis Tech

&

Book 2
Cryptography
By Solis Tech

&

Book 3
Open Source
By Solis Tech

Book1
Circuit Engineering
By Solis Tech

The Beginner's Guide to Electronic Circuits, Semi-Conductors, Circuit Boards, and Basic Electronics

Circuit Engineering: The Beginner's Guide to Electronic Circuits, Semi-Conductors, Circuit Boards and Basic Electronics

Circuit Engineering: The Beginner's Guide to Electronic Circuits, Semi-Conductors, Circuit Boards and Basic Electronics

Table of Contents

Introduction

I want to thank you and congratulate you for purchasing the book, **Circuit Engineering: The Beginner's Guide to Electronic Circuits, Semi-Conductors, Circuit Boards, and Basic Electronics**.

This book contains a beginner's course on circuit engineering. Here, the basics of electric and electronic circuits are discussed. You will grasp the definitions of circuits, semi-conductors, resistors, inductors, transformers, circuit boards, and electronics, in general. You'll even be introduced to electrical safety tips and a set of skills needed in electronics, as well as a short take on reverse engineering, hacking, microcontroller programming, and robotics.

Alongside, you can apply all that you'll be learning once you get started with the proposed circuit projects for beginner. You'll also be rewarded a peek at different career-advancement possibilities. While reading about the fundamentals and various theories in the subject is important, hands-on learning is equally important. This way, you can put your newly gathered knowledge to good use.

If you're uncertain whether or not you have what it takes to learn the ropes in circuit engineering, let this book help you decide. Chances are, you have the stamina for the field and for all you know, you can discover a new passion for circuits and electronic devices.

Thanks again for purchasing this book; I hope you enjoy it!

Chapter I – First Things First: An Introduction to Circuit Engineering

In 1882, there was a *circuit war*; it was between the notable electrical engineers and scientists, *Thomas Edison* (inventor of the DC system) and *Nikola Tesla* (inventor of the AC system).

While Thomas Edison stated that an efficient way of distributing power was via a *DC system*, Nikola Tesla argued that although DC systems are efficient, an *alternating current* is the *more practical* option. It started as a simple clash of ideas, but it eventually led to a major rift. Neither professional conceded; both of them insisted that their own systems were "better".

In the end, it was *Nikola Tesla* that took home the *glory*. Case in point? He was granted funds by an internationally recognized firm, Westinghouse. The majority of the power sources of New York City were based on the ideas of the Serbian engineer; at Niagara Falls in Canada, a power plant was built.

If you're interested in finding out more about the particular *circuit war, AC and DC systems*, and all critical discussions on circuits, taking a course about electronic circuits is the way to go.

I.A. - What Is a Circuit?

Both an *electric circuit* and an *electronic circuit* refer to a complete pathway for electric current, which starts and ends at a single point; it is a passage that allows the electricity to enter at one place, then, let it pass through a series of stops, and finally, leave it to exit at the same place. The list of basic examples of a circuit includes a *light switch* (off and on) and *battery-operated lamps.*

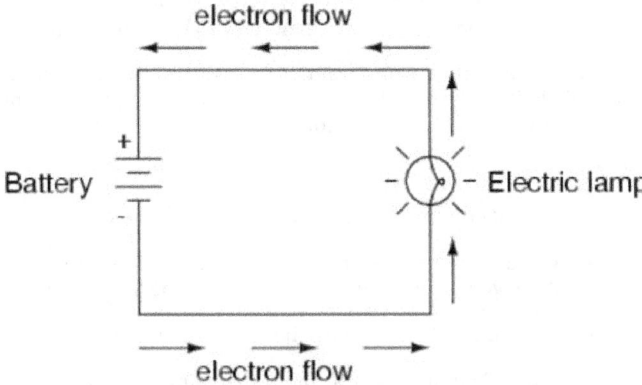

A circuit that follows a fundamental design

A circuit can function well - granted that its design is well-conceptualized. As much as possible, it is recommended that arriving at a simplistic product should be the goal; the simple and straightforward a design is, the better. With a fundamental concept, even if other (beginner-level) circuit engineers who will subject it to inspection will not have a difficult time in understanding its flow. Although there may be complex systems, the agenda is not intended to complicate the explanations.

Moreover, a circuit can be referred to as a space with a conductive path that grants electrons the opportunity to move freely. To create one with a brilliant design, a tip is to learn about the classifications of all circuits. You can use the knowledge to determine the appropriate kind of network, as well as the need for an external or internal source.

2 classifications of a circuit:

1. Linear or non-linear – a circuit that is based on either linear or non-linear networks; it is composed of independent and/or dependent sources and passive elements

2. Active or passive – a circuit that is based on either the absence (passive circuit) or the presence (active circuit) of a source; a source can be a power source or voltage source

I.B. - A Circuit & Its Types

Not all circuits are alike. In fact, one of the most common misconceptions involves an *electric circuit* and an *electronic circuit*; both are said to be one and the same, but they are not. While the former can carry *average to high voltage*, the latter has the tendency to have *low voltage load*.

Moreover, it is always important to be aware of the different circuit types, especially if you're about to make your own circuit; the kind of circuit that you create needs to have the ability to handle a preferred load.

Circuit types:

- Closed circuit – it is a circuit that is fully functional

- Open circuit – it is a circuit that can no longer function due to a damaged or missing component, or a loose connection

- Short circuit – it is a circuit that comes without a load

- Parallel circuit – it is a circuit that connects to other circuits; it is like the main power source or the primary circuit in a series of circuits

- Series circuit – it is a circuit that connects to other circuits; the same amount of electricity is distributed to each of its component circuits; the main power source or the primary circuit is unclear

I.C. - Conductors, Insulators & Semi-Conductors

Conductors, insulators, and *semi-conductors* give light to the fact that a circuit's electrical properties are dependent on the circuit type, as well as on their conduction bands (i.e. their allowed electric power). For instance, if a particular power source chooses to distribute a 9-volt electric power to a closed circuit, its electrical properties can be evaluated by using 2 details: (1) its characteristic as a closed circuit and (2) 9-volt electric power.

Moreover, conductors, insulators, and semi-conductors are integral concepts to the *conductivity* of an object. While conductors and semi-conductors are grouped to describe *charged carriers*, insulators are still considered as relative despite not containing any free charge.

Circuit Engineering: The Beginner's Guide to Electronic Circuits, Semi-Conductors, Circuit Boards and Basic Electronics

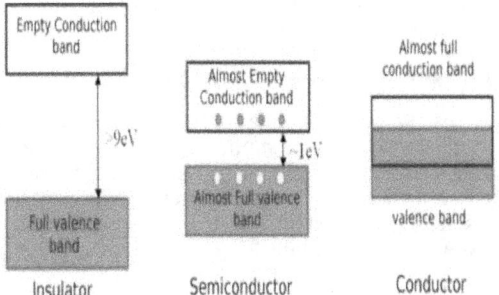

Insulators can be any "ion-less" object; the most common examples of semi-conductors are copper and aluminum & for conductors, gold and silver

Conduction bands:

- Conductor – it is a conduction band that is referred to as the almost full band

- Insulator – it is a conduction band that is referred to as an empty band

- Semi-conductor –it is a conduction band that is referred to as an almost empty band

I.D. - Breaking Down the Components of a Circuit

A circuit can be either *simple or complex*, and be both *simple and complex*. If the circuit in subject is a series circuit, with a group of 10 different circuits that are connected to it or if the said circuit is just a basic closed circuit with 5 different stops, it can be rather confusing to trace. However, if you dissect any circuit, you'll discover 3 *constant,* integral components.

Integral components:

1. Load – it is the representation of the power consumption, as well as the work that is accomplished within a system; without it, there's barely a point in having a circuit

2. Power source – it is where the electricity comes from

3. Pathway – it is the framework of a circuit; from the power source, it follows the load through each of the network, and finally returns to

and exits the power source; it is also referred to as the conductive pathway

I.E. - The Roles of Current, Resistance & Voltage

Current, resistance, and voltage are the 3 representations of the important components of a circuit's system. They can explain how electricity enters, then, moves from 1 point to another, and finally exits. Whether the path of electricity is rather simple, these representations remain constant. Apart from describing the electric flow, they can serve as indications of faults (in instances when a circuit fails to work).

3 representations:

- Current – it is the representation of the electric flow; particularly, its focus is on the flow of electrons

- Resistance – it is the representation of the nature of an electric flow as it moves around the circuit

- Voltage – it is the representation of the electric force or pressure; in general, the supply comes from an electric outlet or a battery

I.F. - AC/DC Systems: Which System Is in?

AC and DC systems (or alternating current and direct current systems) are often associated to each other. When the *AC system* is mentioned, so is the *DC system*. Conversely, when it's the DC system's turn to be in the spotlight, it won't be long until the AC system is mentioned. This is because these systems are opposite of one another; to get a better understanding of one of them, it's recommended to be familiar with the other, as well.

Moreover, *AC and DC systems* are types of a circuit's current flow. In an AC system, the current flow changes its direction occasionally. Meanwhile, in a DC system, the current flow follows a single direction.

It can be deduced, therefore, that an AC system grants a circuit freedom to let the current flow in several directions. While this can be an advantage, this doesn't permit the continuous flow that a DC system can entitle.

So, should you use an *AC system* or a *DC system*? The decision as to which current system is dependent on the more practical design to follow; take into account the aim of having your own circuit. If you prefer something grand and you intend to power something large, the AC system can step in. On the other hand, if you're good with a basic setup, you can use the DC system's concept as basis.

I.G. – What Is a Transformer?

A *transformer* is a device that serves as a portal for energy transfer within the points in a circuit or from circuit 1 to circuit 2. In most cases, it is used for increasing and decreasing the voltages in a system.

When the first transformer was built in the mid 1880s, circuit engineers discovered that a transformer significantly improves the electric flow in a circuit, and consequently, results to a more powerful circuit. The discovery made way for various transformer designs, as well as various transformer sizes.

A primary principle of a transformer is its need for extremely *high magnetic permeability*. It follows that a circuit that is capable of attracting power is, of course, more inclined to have electric current transferred to it; and, conversely, a circuit with *low magnetic permeability* is less likely to extract power from another circuit.

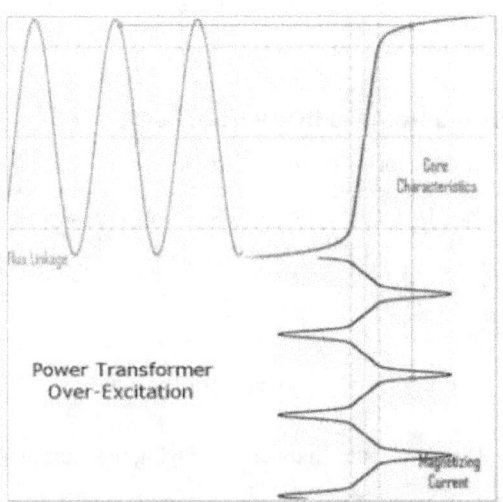

Magnetic permeability is defined as the ability of a circuit to hold and support an internal magnetic field

Chapter II – The Anatomy of a Circuit

Let's say that a lamp in your room has been around for a while, and let's say it chose to give up on you the other night; although it's an old lamp, you still believe it can *work fine*. You suspect that the problem is in the switching board. Instead of just letting it be and since you're interested in how it works, you choose to study its components.

As you open its internal system, you see that connected to some sort of panel are 2 wires; one wire is red, the other is black. As far as your knowledge in circuits can tell you, the 2 wires in your hands are: (1) a live wire or the wire that is connected to a switch, and (2) a neutral wire or the wire that carries the load.

As the universal rule in circuit engineering goes, *red* is the color that indicates a live wire, *black*, on the other hand, is a neutral wire. For a chance to know why your lamp gave up on you other than old age, checking the red wire would be a good start.

II.A. – Individually Speaking: Parts of a Circuit

A circuit can *work smoothly* if its individual parts are contributing to the workflow as expected. Remember that it follows a series; if one point in that series is not in condition. So, if you're wondering why a system is functional, check each one of its components. To use the old adage (and modify it a bit), the circuit, as a whole, is just as good as its individual parts.

Individual parts:

- Capacitor – it is in charge of the stability in the power source in a circuit
- Diode – it is in charge of supplying the light in a circuit
- Inductor – it is a coil of wiring system
- Resistor – it is in charge of power consumption
- Transistor – it is in charge of the electric signal control

II.B. – Circuit Categories: Which Category Do You Belong to?

Circuit categories describe the *voltage levels*, as well as their *electric flow over time*; they refer to how much and how powerful the current is as it enters and passes through each point in a circuit. Since they vary, it's listed as an effective way of understanding circuit systems of sorts if they are categorized.

Circuit categories:

- Analog circuits – these are the categories of circuits that use the concepts of parallel circuits and series circuits as basis. Among their fundamental parts are capacitors, diodes, resistors, and wires.

  In diagrams, analog circuits are easy to recognize. Usually, when a model of the particular circuit is drawn, a simple illustration is presented since these circuits do not follow a complex system. In most cases, when illustrated, the parts (e.g. capacitors, diodes, wires, etc.) are represented by lines.

- Digital circuits – these are the circuits that rely heavily on Boolean algebra (i.e. values are either true or false, or as denoted 1 or 0); therefore, these circuits are often dependent on transistors that can create closed logic. Compared to analog circuits, these follow a state-of-the-art design.

  Furthermore, digital circuits are designed to create either numerical or logical values for the representation of electricity that flows within their system. Since these are not only focused on the mere ability to take in electric current, but rather, on the individual properties of all of their parts, as well, these circuits come with advanced functions; they can provide memory and accomplish arbitrary computations.

- Analog-digital – these circuits are sometimes called hybrid circuits or mixed signal circuits since both system designs of analog and digital circuits are given light. Although their concept can be quite complex, these circuits can deliver a more thorough result; the procedures are combined, which allows collaborative effort from different parts. One example is a telephone receiver; first, it works based on analog circuitry to create and stabilize signals, then, based on digital circuitry, these signals are converted into digital units, and finally, are subjected to interpretation.

II.C. – Where Does Inductance Enter the Picture?

In circuit analysis, the term *inductance*, introduced in 1886 by *Oliver Heaviside*, refers to the property of a circuit's electricity-producing component to change in amount. Apart from the support of a circuit's aspect to vary, it points out the need for a filter and energy storage systems to be provided.

As it follows, the component in a circuit that enables inductance is called an inductor. Usually, these parts are made out of wire. But, while some circuits contain inductors as integral parts, others remain functional without the need to alter the electric flow.

Inductance can be either *mutual inductance* or *self-inductance*. The former refers to a change in electric current from one inductor to another inductor; it explains the primary operations of a transformer. Meanwhile, the latter refers to the *stable inductance* within a system.

Moreover, inductance is represented by the symbol *L,* which is meant to giver to the scientist, *Heinrich Lenz.* It also measured in units of *henry* after the American scientist, *Joseph Henry*; it follows that although it was Oliver Heaviside who introduced the term, the man behind the development is Joseph Henry.

Mutual inductance describes the occurrence in a circuit when there is change that can be traced to an inductor; particularly, it refers to the alteration due to an inductor's preference of a nearby inductor. It is essential to learn about the relationship of *2 inductors* since it is the basis of the operations of a transformer. Additionally, a limit has to be maintained in order to keep potential energy transfers regulated; the failure to incorrectly calculate mutual inductance can result to *unwanted inductance coupling*, as well as a *power overload*.

It can, therefore, be deduced that mutual inductance (as represented by the symbol *M*) is the measurement of the coupling that involves 2 inductors; the 2 inductors are, then, given particular importance (in terms of coil turns), along with each inductor's ability to admit current flow or *permeance*. The formula for calculating mutual inductance is as follows:

**representations: M_{AB} = mutual inductance in circuit A and circuit B

N_B = inductance in circuit B

N_A = inductance in circuit A

P_{AB} = permeability in circuit A and circuit B

$M_{AB} = (N_B)(N_A)(P_{AB})$

When explained, the formula highlights that the mutual inductance between inductor A and inductor B (or M_{AB}) is equal to the product of 3 elements: [1] the coils of inductor B (or N_B), [2] the coils of inductor A (or N_A), and [3] the permeance of inductor A and inductor B (or P_{AB}).

On the other hand, self-inductance refers to voltage induction of a current-carrying system in respect to the changing current in the circuit. It points out that eventually, there will be another current that will flow along with the primary current. Due to the amount of force within the magnetic field, voltage is induced; particularly, voltage is *self-induced*.

II.D. - What Makes an Integrated Circuit?

An *integrated circuit* (or IC) is alternatively called a microchip or a chip, due to its size. It works depending on a particular signal level. One example is the integrated circuit that enables a computer to perform a multitude of tasks; instead of loading a computer's structure with a large circuit, it comes to the rescue.

In most cases, an *integrated circuit* operates at little defined states. Compared to the normal circuit whose operations are distributed over continuous amplitudes, it can function within a small network; the normal circuit may sometimes fail to work with only minor amplitude ranges.

Basically, an *integrated circuit* is no different from any other circuit; its power can astound you, yes, but, if it comes down to describing how it is, it's simply a circuit that has been reduced so it can fit inside a chip.

Chapter III – Resistance Isn't Futile

Without a material that can act as the opposing force, a circuit can *function*, but it may not function *as desired*. When an electric supply can perform its function by distributing electricity to the opening of a circuit, the electric current will keep on flowing; its flow can be uncontrollable, which can destroy a system's integrity. Usually, without the opposition, a circuit ends up taking too much load.

The term for this opposing material is *resistance*; it goes hand in hand with the term conductance. And, as mentioned in the first chapter, it is the representation of the current flow in a circuit.

III.A. – What Is Resistance?

Resistance is the measurement of an opposing electric current; it can be expressed in ohms. It generates an amount of friction that is relative to the necessary amount of electricity that a particular circuit can handle.

In a way, *resistance* is responsible for the *smooth flow of electricity* in a circuit. Although others would counter the argument by saying that rather than support the effortless flow of electricity in a system, it slows it down.

However, it is *resistance* that allows *balance of electricity* in a circuit. Take for example the case of a circuit that can only handle a total of 15V. If a circuit takes in 20V at a resistance of 5 ohms, the number is diminished to 15V, which indicates a functional circuit. Conversely, if, in the same situation, there is no *resistance* of 5 ohms, a circuit may not be as functional as desired; its system ends up carrying 20V, which implies that it is overloaded.

III.B. – Resistive Circuit 101

A *resistive circuit* is a kind of circuit that consists of nothing but a series of resistors to complete the combo of electric current and voltage source. If viewed in a chart, it is noticeable that the power waveform is always positive; it is suggestive the power in a circuit is always dissipated, and is never returned to the original source.

It is important to note that the frequency of the power in a circuit should not be equal to the frequency of the electric current and voltage. If possible, the frequency of the power should be twice as high as that of the electric current and voltage. This unequal frequency distribution grants constant change within a system.

Since it is made up of resistors and does not include transistors and capacitors, a resistive circuit is rather easier to analyze. Understanding the electric flow within the circuit (whether in an AC or DC system) requires a

16

straightforward technique. Therefore, determining the flow of the current in a resistive circuit is simple; by adhering to the formula, calculating the figure is easy.

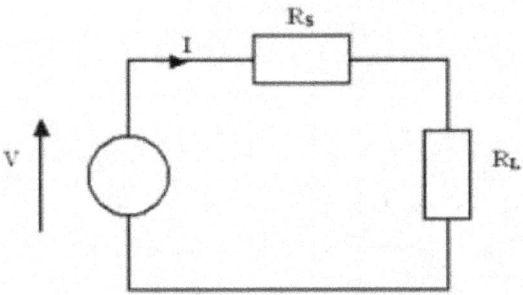

In a resistive circuit, voltage can easily be monitored

**representations: I = total current

RS = Resistance Source

RL = Resistance Load

I = voltage ÷ (RS +RL)

III.C. - War between the Types of Resistance

Resistance is classified according to the type of resistivity that it can contain, along with the amount of resistance that a circuit can carry. This allows the opportunity for an opposing force to be valued, regardless of its resistivity. As the professional electrical engineers can attest, not every circuit component that produces resistance satisfies the rules, particularly, *Ohm's Law*.

2 types of resistance:

- Differential resistance – it is the resistance derivative of voltage in light with the electric current; also referred to as *incremental resistance, small signal resistance, or dynamic resistance*, its concept is responsible for oscillators and amplifiers

- Static resistance – it is the resistance that corresponds to the typical definition of resistance; it is also called *chordal resistance* or *DC resistance*

III.D. - Resistance vs. Conductance

The average circuit comes with both *resistance* and *conductance*, which gives balance to the electric flow in a circuit's system. While the former refers to the opposition, conductance describes the amount of current that is converted into power that revolves around different points.

Conductance also covers the ability of a circuit's components to conduct electricity. And, to bring light to its counterpart's ability to oppose the flow, it dwells on the subject of the convenience of electricity to pass through a series of points in a circuit.

With both the resistance and the conductance in the system, a circuit can function as desired.

III.E. – The Need for Calculations (Four Ways)

For electric current to flow smoothly within a system, a level of resistance has to be present. And, since not all circuits come with a similar design, their resistance levels vary. To calculate a particular circuit's resistance, first, you need determine its type, and its provided values, as well.

Four ways:

#1 – Resistance calculation for a series circuit

The formula:

Resistance = $_1R + _2R + _3R + _4R$

#2 – Resistance calculation according to voltage & power

**representations: total voltage; PT = total power

The formula:

Resistance = $VT^2 \div PT^2$

3 – Resistance calculation according to voltage & current

**representations: VT = total voltage; IT = total current

The formula:

18

Resistance = VT ÷ IT

\# 4 – Resistance calculation according to power & current

**representations: PT = total power; IT = total current

The formula:

Resistance = PT ÷ IT

III.F. – What about Sheet Resistance?

Sheet resistance refers to the measurement of the resistance in a thin sheet in a circuit's components. It can be used to describe the resistibility of different circuits and can point out the specific difference in circuits that vary in size. Especially in the case of a commercial product, the topic is covered for the assurance of quality.

You can look at *sheet resistance* as a special kind of resistance since it generates a more specific value. Usually, the average resistance in a circuit is expressed in *ohms*; *sheet resistance* is expressed in *Ohms per square*.

In most cases, *sheet resistance* is used for the analysis of circuits with uniform conductivity or semi-conductivity. Typical applications are extended to quality assurance for a commercial circuit.

III.G. – The Role of Impedance & Admittance

Like resistance, *impedance* can be described as the opposition in a circuit; unlike resistance, however, it refers to the opposing force of a circuit after the application of voltage. It is only relevant to AC systems or circuits where direct current isn't the supplied.

It was in 1893 when the concept of *impedance* was initially introduced by the Irish engineer, Arthur Kenelly. Back then, it was denoted by *Z* and is defined as a complex number.

When it comes to quantitative terms, *impedance* refers to the ratio of voltage to the electric current in a circuit. Its introduction is important for beginners especially if they're scratching their heads as to why there's an opposing force besides resistance.

Impedance, like resistance, comes with values. In a single open circuit, its value is presented in *ohms*. In the event of a series circuit or a parallel circuit, its value can be calculated by simply adding all the defined values in each unit.

The formula:

**representations: TZ = total impedance

Z_1 = impedance in component 1

Z_2 = impedance in component 2

Z_3 = impedance in component 3

Z_{10} = impedance in final component of a circuit

$$TZ = Z_1 + Z_2 + Z_3 \dots Z_n$$

Meanwhile, *admittance* is a relative concept in circuit engineering. It addresses the issue that alongside the difference in the magnitude of the electric current and voltage that are flowing within a circuit, the difference in phases needs to be given light, too. This way, the maximum load within a system can be calculated accordingly.

The formula:

**representations: TY = total admittance

Y_1 = admittance in component 1

Y_2 = admittance in component 2

Y_3 = admittance in component 3

Y_{10} = admittance in final component of a circuit

$$TY = Y_1 + Y_2 + Y_3 \dots Y_n$$

Chapter IV – It's Time to Measure the Electric Flow in a Circuit

In the previous chapter, the formulas for the calculation of a circuit's resistance levels were shared. However, the formulas for the *calculation of the entire load* in a circuit have yet to be discussed.

This is due to the significance of using the appropriate measurement units. In a few cases, especially those who are still on the initial phase of learning circuits? They aren't quite careful with their selected units. As it follows, it's not only necessary to calculate a circuit's electric flow; it's also necessary to calculate a circuit's flow correctly.

IV.A. – Standard Units

Among the several reasons to use *standard units of measurement* are for indications of exact measurements and for indications of the preferred measurements in a system. These units bring uniformity.

In circuits, the usual standard units that you encounter are *V, W, I,* and *P.* Although there are more, those who wish to explain a system's electric flow rely on these measurements; rather than introduce a bunch, which may only make matters more confusing, some are preferred. Moreover, without such units, understanding others' discussions of circuits is nearly impossible.

Standard units:

- Conductance – its measuring unit is Siemen with G as symbol

- Current – its measuring unit is ampere with I or i as symbol

- Frequency – its measuring unit is Hertz with Hz as symbol

- Inductance – its measuring unit is Henry with H or L as symbol

- Power – its measuring is watts with W as symbol

- Resistance – its measuring unit is ohm with R as symbol

- Voltage – its measuring unit is volt with V or E as symbol

IV.B. – Commonly Used Alternatives

Other than the *standard units of measurement*, other units are given light since these can enable clearer expression of the electric flow in a circuit. Especially if the circuit in subject contains a rather complex system, it can be difficult to arrive at a definite solution.

Other units:

- Angular frequency – it is a unit of measurement used in an AC circuit; it is a rotational unit that describes the relationship of at least 2 electric forms in a circuit

- Decibel – it is a unit of measurement that represent the gain in either current, power, or voltage; since it is only a tenth of the original unit, *Bel*, it is primarily reserved for denoting extremely small amounts

- Time constant – it is a unit of measurement that describes the output of a circuit's minimum or maximum output value; in a way, it refers to the measurement of time reaction

- Watt-hour – it is the unit of measurement that describes the electrical energy consumption over a period of time

IV.C. – Units of Force

Since *force* in a circuit is an important concept, it is advised that the particular unit of measurement is presented correctly. Even in physics, it is reiterated that it should be labeled according to its right category.

Atomic and electrostatic units of force:

- Hartrees

- Newtons

- Tesla

- Coulombs

- Meters

Chapter V – Power Transfer at Max

There will be instances of a *power transfer* in a circuit. For the electricity to continue its smooth flow, its original energy source will be replaced with an internal energy source. With such a change (especially in the case of a series circuit where power needs to flow continuously), the explanation for its system's pattern becomes a notch challenging.

For *beginners*, a great way of understanding *power transfer* in a circuit is to understand maximum power transfer, along with the concepts that dwell on the topic. As it follows, by gaining clarity on how much was the original power, as well as how much power a particular circuit can handle, you can see whether a power transfer is necessary or will only cause its load to be compromised.

V.A. – Maximum Power Transfer

Maximum power transfer, a concept that was introduced by Moritz von Jacobi sometime in the 1840s, draws light on the idea that for maximum external power to be obtained, an internal resistance needs to be in place. However, the transfer can only be flawless if the original resistance is equal to the potential power that an internal resistor can produce.

Consequently, maximum power transfer yields results that point out *power transfer*, and not *efficiency*; while improved efficiency can be a byproduct, it is not the chief purpose of maximum power transfer. It implies that although higher percentage of power is transferrable, it does not affect the magnitude of the power load (i.e. the extent that it can affect a circuit). In the event that the internal resistance is modified to accommodate a value higher than the value of the original resistance, improved efficiency can be achieved.

Moreover, the concept of maximum power transfer was initially misunderstood; a subject of many arguments was a circuit's reduced efficiency with the occurrence of transfer. Some insisted that due to the potential power that is lost during an exchange, a circuit may fail to reach 100% efficiency. As emphasis of this group's angle, take for example the case of a motor whose power is transferred from a battery; power in this situation may not be maximized, and it will only be realized over time when battery power has been fully consumed.

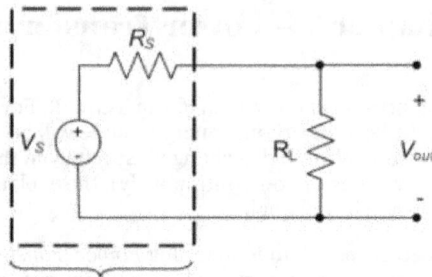

The maximum power theory states that the task of transferring power consumes power, too

It was *Thomas Edison*, as well as his fellow scientists, *Francis Robbins Upton*, who contested that maximum power transfer and efficiency are only *relative*; the 2 concepts are not one and the same. In fact, there is a discussion about *maximum power efficiency*, too.

In the exchange, you will find that *resistance* plays an important role. By giving light to the former argument, you can calculate a circuit's capability of a maximum power transfer, in relation to maximum power efficiency with the following formula:

MPT = RL ÷ (RL + RS)

A circuit's *MPT* (or maximum power transfer) can be determined with basic arithmetic skills. First, divide the *RL* (or Resistance Load) by the sum of the *RL* and *RS* (or Resistance Source).

V.B. – Thevenin's Theory

Thevenin's Theory, conceptualized by Hermann von Helmholtz and Leon Charles Thevenin, discusses that if a circuit follows a linear network, any point can be replaced given that it remains to carry a source for current, resistance, and voltage. Behind it, the idea is to supply an equivalent.

Originally, *Thevenin's Theory* can only be applied to circuits that operate with a DC system; since a DC system is rather simple, replacing its components with an equivalent is possible. Eventually, however, its capability to handle a load in a non-linear system was discovered; it can offer solutions for an AC system.

Moreover, the *Thevenin Theory* puts emphasis on that the average circuit can only be considered to have a linear according to a limited range; it can only be replaced by the components of with values among the range.

The Thevenin Theory follows that power dissipation can yield unique values, and can also yield identical values. However, the results can only be accomplished with the power supplied by an external resistor.

V.C. – The Star Delta Transformation

The *Star Delta Transformation* dwells on the idea that a circuit's system can change from one phase to another. For instance, if a circuit's power source is altered, its ability of carrying power from a point to the next is altered, too. Especially if there are 3 branches in a circuit's system, the power that circulates is known to form a closed loop.

The Star Delta Transformation refers to 2 kinds of circuit transformations. The first circuit transformation is a star transformation, which can be described by a "Y" formation; the second circuit transformation is a delta transformation, which can be described by a triangular pattern.

Moreover, the Star-Delta Transformation describes a 3-phase network of circuits, which can explain power transfer between these 3 networks. It enables the conversion of impedances that are connected to each other. With the theory as basis, alongside getting a clear scope for power transfer analysis, solving various concerns can be accomplished, too; the concept is applicable to different types of circuits including *series circuits, bridge-type networks, resistive circuits,* and *parallel circuits.*

The Star-Delta Transformation can be converted to the Delta-Star Transformation. From the star or Y-formation, the circuit creates a triangular network as the transition is achieved.

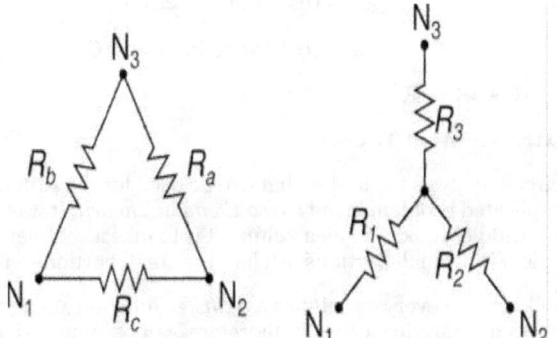

The Star-Delta Transformation or the Delta-Star Transformation is also called the Y-Δ Transformation or the Δ-Y Transformation

For the transition of the Star-Delta Transformation into the Delta-Star Transformation, a formula should be followed; this is meant to ensure that the transformation, along with the calculations for the total resistance in *all 3 circuits*, is successful. Initially, the goal is to compare the amount of power in an individual network. Once the power in network 1 has been acknowledged, proceed to identifying the weight that one network holds in the entire formation; one way of determining this is to disconnect that entire network and observe the operations of a circuit.

The formula:

**representations: $_\Delta R$ =   total resistance of the transformation

$_1N$ = resistance in node 1

$_2N$ = resistance in node 2

$_AR$ = resistance in circuit A

$_BR$ = resistance in circuit B

$_CR$ = resistance in circuit C

$$_\Delta R \,(_1N \,_2N) = {}_CR \,||\, (_AR + {}_BR)$$

The simple version of the formula:

**representation: TR = total resistance

$_AR$ = resistance in circuit A

$_BR$ = resistance in circuit B

$_CR$ = resistance in circuit C

$$_TR = {}_AR + {}_BR + {}_CR$$

V.D. – Extra Element Theory

A circuit analysis technique that can be used for the simplification of a complicated problem is *the Extra Element Theory*; it was proposed by R.D. Middlebrook. The idea behind it is to take a complex matter, then divide it into small portions; each of the small portions will be addressed.

It follows that every circuit has a *transfer function* and *driving point*; the process of analyzing a circuit, therefore, can become easier if the aforementioned elements are first identified.

In the Extra Element Theory, unlike in other circuitry theorems, an element such as a capacitor or resistor can be temporarily removed so the transfer function or driving point can be determined. Since there are circuit components that can complicate an equation (regardless of how integral they are to a circuit), it is practical to set them aside for a while; although they may be of value to a circuit as a whole, it was proven that they don't affect calculations. Once the initial goal is achieved, the elements can be returned.

Impedance is a familiar term in discussions of the Extra Element Theory; it can be analyzed with the employment of the theory. In certain cases, its input can be determined in network granted that an *extra element* joins in.

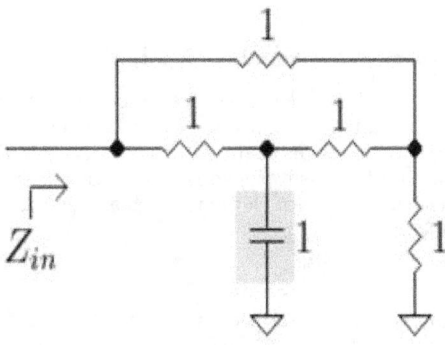

The Extra Element Theory (in relation to input impedance) proposes the addition of an "extra" element that is equivalent in value to the other elements

The formula for finding the impedance is:

**representations: Z = impedance

s = source

$Z = 1 \div s$

V.E. – Simplification of the Source

Simplification of the Source, sometimes referred to as Source Transformation, is the process of converting electric current into voltage, or voltage into electric current. It is a common technique used by many circuit engineers for explaining their circuit's system in simple terms.

The process of Simplification of the Source usually begins with an existing resistance source in a circuit; it is then replaced with new electric current source with a similar level of resistance. Since it is *bilateral procedure*, one can be derived to yield results from another. It makes way for the adjustment of voltage as it gradually becomes the equivalent of a particular circuit's resistance.

Moreover, *Simplification of the Source* may begin with an existing resistance, but is *not limited to the accommodation of resistive circuits*. It means that the process can be performed on circuits that involve inductors and capacitors.

V.F. – Where Does the Rosenstark Method Fit in?

The *Rosenstark Method*, sometimes called *Asymptotic Gain Model*, is yet another important subject where power transfer is concerned. In light of the return ratio, it serves as the representation of *negative gain* from feedback amplifiers. As it provides an intuitive form of circuit analysis, it introduces a new batch of elements such as the *return ratio* and *asymptotic gains*.

The Rosenstark formula:

**representations: G_o = o asymptotic gain

G_∞ = infinite asymptotic gain

T = return ratio

Rosenstark Method = $\{G_o + [T \div (T + 1)]\} + \{G_\infty + [T \div (T + 1)]\}$

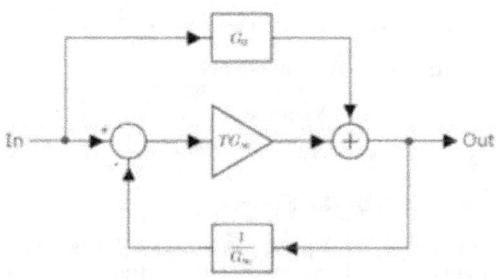

Circuit Engineering: The Beginner's Guide to Electronic Circuits, Semi-Conductors, Circuit Boards and Basic Electronics

The Rosenstark Method serves as the basic representation of power transfer in a circuit

Best features of the Rosenstark Method:

- Assumes that the direct transmission is small and can be equated to the asymptotic gain

- Characterizes bilateral properties and feedback amplifiers

- Identifies (via thorough inspection) the passive circuit elements

Steps:

1. Choose a source in a circuit's system; preferably, opt for a dependent source.

2. Determine the source's return ratio.

3. Identify the G_∞.

4. Identify the G_0.

5. Use the Rosenstark formula and substitute values.

Chapter VI – Laws, Laws & More Laws

Come to think of it, *circuit analysis* is one of the broad branches of electrical engineering. It covers the basics of circuits and power consumption; it also covers the extensive aspects of the topic such as the operations of an entire electrical network.

Without laws regarding how electric current is distributed within a circuit's components, there is a risk of unclear discussions of *fundamental and in-depth analysis*; without them, a guide that can break down lengthy components is absent. For beginners, especially, they are important since they grant the chance to understand the functions of a circuit and even each one of its parts. It stays true to the idea that for the resolution of a big problem, looking at it like little pieces is an opportunity to conqueror its difficulties.

VI.A. – Putting Kirchhoff's Laws into Motion

Kirchhoff's Laws are composed of *2 equations* that dwell on the conservation of charge and energy. They were initially introduced in 1845 by Gustav Kirchhoff as a means for the determination of a circuit's power consumption, as well as its parameters.

As several discussions go, Kirchhoff's Laws may not be too important since it was derived from the work of Scottish physicist, *James Maxwell*; since it was derived from another's work, it is said that a circuit engineer should rather use the original work as basis. However, it is a set of laws that primarily focuses on the operations of a closed circuit; the previous work from which it was taken doesn't put emphasis on a closed circuit, and is rather descriptive of the generic circuit.

Moreover, Kirchoff's Laws are conditional. It may be useful in describing charge and energy conservation for different circuit elements, but it can only yield an *approximation*; it also requires certain factors such as changing electric currents, voltage, and resistance.

Kirchhoff's Law is actually a general term. To be specific, it addresses 2 subjects: current and voltage. It acknowledges all circuits, regardless of complexities.

To solidify its argument in the conservation of charge and energy, one of Kirchhoff's Laws, *KCL* (or Kirchhoff's Current Law), states that the electric current in an interconnected network is relative. And, to emphasize his point, Gustav Kirchhoff contests that the algebraic sum of the electric current in a joint network is *0*.

Another law of Kirchhoff that complements the KCL is *KVL* (or Kirchhoff's Voltage Law). Its basis is on the general law on energy conservation that defines voltage as the energy per unit of charge. Just like in the previous law, Gustav Kirchhoff says that in a closed network, the algebraic sum of the electric voltage in a joint network is *0*.

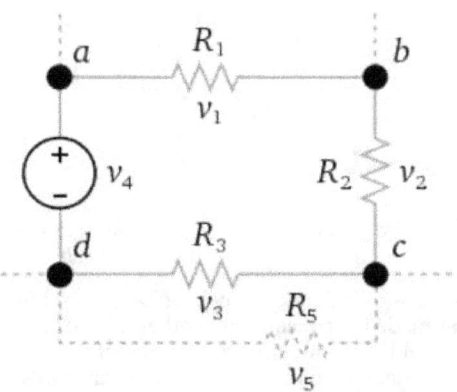

According to KVL, adding all the voltages within the circuit yields 0

VI.B. – Does Michael Faraday Know What He's Getting Into?

The *Law of Induction* by *Michael Faraday* is a fundamental law that revolves around the field of a circuit and the possible elements that are subject to eventual interaction. It carries both qualitative and quantitative aspects, and proposes that only with the presence of an infinite source of inductors can a circuit retain its inductive capability.

Unfortunately, not all circuit engineers and scientists are one with *Michael Faraday* and his *Law of Induction*. It is argued that although it holds some matters true, especially where loops of wire are concerned, it can yield the wrong results if used extensively; it can only handle a certain field and is usually arbitrarily small. In fact, counterexamples were previously presented.

A counterexample of Faraday's Law of Induction is the case involving an *electric disc generator*. Due to its own magnetic field, it can rotate circularly at a specified angular rate; it can complete a rotation and in the process, induct electricity by distributing it to other areas (within the circuit). It can be deduced that although the circuit's shape has remained constant over a period of time and although it can induct electricity, its distributive method made it lose inductive capabilities.

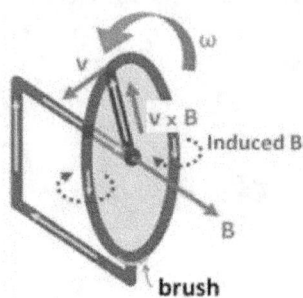

An electric disc generator induces electricity until cycle completion in the object's lower brush.

Faraday's Law of Induction may have been noted for certain flaws, but it is a law that made the conception of other important laws (not all of them are in the field of circuit engineering) possible. An important law with basis on Michael Faraday's idea is Albert Einstein's *Special Relativity*.

VI.C. - Ohm's Law & Conductivity

Ohm's Law, named after the German scientist, Georg Simon Ohm, is a powerful law that describes the resistance and electric current in a circuit, as well as their potential difference. Particularly, it sheds light on 2 points, then, subjects them to further analysis. Consequently, the relationship between Point A and Point B will indicate each of their properties as individual components.

In the verge of learning its concepts, it will dawn on you that *Ohm's Law* is an *empirical matter*. After observations with different ranges of length scales, it was proposed that it would fail where small wires are concerned; however, this proved to be a mere assumption since it was discovered that it could function regardless of a wire's size.

Since it is a basic equation in circuit engineering, *Ohm's Law* can also be applied for the determination of metal conductivity; it can be relied on when it comes to understanding the electric flow and conductive aspect in a circuit's component. Various materials that make use of Georg Ohm's principles are referred to as *ohmic*.

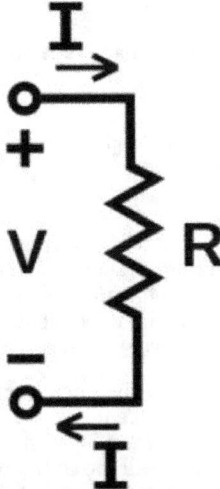

Georg Ohm set the parameters for his law to include only common terms in circuitry

In the physics of electricity, Ohm's Law is important where quantitative measurements in a circuit are concerned. Initially, when it was presented to German scientists, it was only mocked and rejected due to the mere fact that it went against the basic understanding of electric flow; it was dismissed as "a web of fancies" and worse, Georg Ohm was dubbed as a "fancy but invaluable" science professor. It was only until 1840 when it earned recognition and is now widely used today.

How to calculate for the elements in Ohm's Law:

Formula # 1:

voltage = current* resistance

Formula # 2:

current = voltage ÷ resistance

Formula # 3:

resistance = voltage ÷ current

VI.D. – The Argument of Norton's Theorem

Edward Lawry Norton, as he implies in *Norton's Theorem*, proposed that in circuit analysis, any linear network, granted that it has sources of voltage, electric current, and resistance, has an equivalent in dual network. A certain system may be unique, but values that can make a similar model can be obtained.

To find a circuit's equivalent using Norton's concepts, certain values need to be identified: voltage, electric current, and resistance. Once there are clear sources, you can begin formulating the necessary equations. However, when the sources are dependent on each other, another (and it has to be from a generic source) electric current needs to enter the picture.

Moreover, Norton's Theorem gives light to the fact that the values for the *equivalent current* and *equivalent voltage* are values that can be identified at the first 2 terminals of a circuit. It is important to note, however, that such a case is only possible if all the components in the said circuit are *short-circuited*. If the components are not short-circuited, a practical solution is to have current and voltage sources replaced; current can be replaced by transforming a circuit into an open circuit, and voltage can be replaced by transforming a circuit into a short circuit.

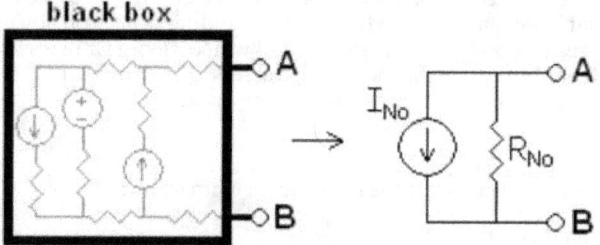

Norton's Theorem was also developed by Hands Ferdinand Mayer

The formula:

**representations: TC_AB = total current in circuit A and circuit B

TC = total current

TC_{AB} = voltage ÷ (TC in circuit A + TC in circuit B)

VI.E. – Coulomb's Law

One of the laws that underwent heavy testing is a law that gives importance to the electrostatic relationship of all the charged elements within a circuit; this law is called *Coulomb's Law* after the scientist, Charles Agustin de Coulomb. It argues that the magnitude of interactive circuit components is in direct proportion to the *squared area* between the distances. To demonstrate a point, it shows that the primary force works along a straight line.

Moreover, Coulomb's Law requires that the placement of the charged elements is accomplished through a single medium. This causes the arrangement to eliminate any possible complications.

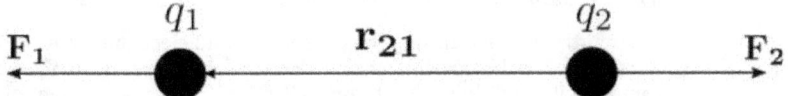

The application of Coulomb's Law is only valid if the circuit components are point charges

To make it effortless to obey Coulomb's Law, the determination of *Coulomb's constant* is recommended. Once the value of the constant has been determined, not only is it easier to adhere to the aforementioned law, you are also ensuring that the calculations for the electric current are valid.

The formula:

**representations: CC = Coulomb's Constant

ε_0 = electric constant

$CC = 1 \div 4\pi \times \varepsilon_0$

Chapter VII – Understanding Electromagnetism

A pair of loudspeakers is an example of an electronic device that works due to the *strategic positioning* of circuit elements. That, along with the integral parts called *electromagnets*; such components cause disturbance to other circuit elements. Their effect on their fellows is rather significant as they tend to make the device's cone to vibrate. As a result, sound, and in most cases, high quality sound, is produced.

For a deeper understanding of a loudspeaker, you could use a lesson in *electromagnetism*.

VII.A. – An Introduction to the Concept behind Electromagnetism

Electromagnetism is a branch of science that involves the study of electricity and magnetism. It follows that wherever there is an electric field, there is also a magnetic field. It was initially introduced and developed by *Hans Christian Orsted*. In 1802, an Italian scholar by the name of *Gian Domenico Romagnosi* also examined the field

In circuitry, understanding the concept behind electromagnetism is important due the possibly strong electromagnetic reaction of circuit components. It follows that the *EMF* (or electromagnetic force) of any circuit serves a major role in the determination of internal properties.

Chapter V, *Power Transfer at Max*, and Chapter VI, *Laws, Laws & More Laws*, contain different discussions about transferring electric current and voltage. In electromagnetism, more discussions about transfer will be tackled. This time, however, the focus is on the charged elements in a circuit (whether negatively or positively charged), as well as their responses when exposed to movement.

The primary principles of electromagnetism:

- The direction of a magnetic field dictates the direction of the current in a circuit

- The electric current of a conductor creates a magnetic field; the formation is dependent on the direction of the conductor, but is always shaped in a corresponding circle

- The magnetic poles of charged elements always come in pairs; one pole in a pair needs to be inversely proportional to the other

- Charged elements can either attract or repel; elements with different charges are attractive to each other, while those with similar charges avoid contact

- Electric current can be induced when it is moved away or toward a magnetic field

VII.B. – Gauss Who?

A notable individual in the field of electromagnetism is *Carl Friedrich Gauss*. He is a German mathematician who put together the critical electromagnetism concepts in the form of a law called *Gauss Law*. According to him, to determine the relevance of the distribution of electric charges, the circuit's electric field, in its entirety, must be evaluated.

Gauss Law allows mathematical expression with the employment of *integral* and *differential calculus*, preferably in vector form. For a beginner in circuitry, the particular technique of demonstration is advised since it grants a clear perspective of the concept of electromagnetism.

The integral formula:

**representations: Φ_E = electric field

ε_0 = electric constant

Q = total electric charge

$\Phi_E = Q \div \varepsilon_0$

The differential formula:

**representations: ∇ = an electric field's divergence

E = the other half of an electric field's divergence

p = electric charge density

ε_0 = electric constant

$\nabla \times E = p \div \varepsilon_0$

VII.C. – The Main Formulas in Electromagnetism

In a circuit's electromagnetic field, it is important to remember that charged elements tend to move radically; sometimes, predicting the direction of the electric flow is nearly impossible. While others go about in a non-linear network, many charges obey the rule of superposition since they adhere to a linear path. With the use of certain laws as basis, the relationship between the charged elements can be evaluated.

4   laws:

1. Ampere's Law – an important law whose applications include instances of a moving magnetic field; particularly, it be applied in situations involving current-carrying wires

   The formula:

   **representations: B = electromagnetic field

   DL = differential element of the electric current

   μo = permeability of o space

   I = electric current in an enclosed circuit

   $\int B \times DL = \mu o \times I$

2. Biot-Savart's Law – a law that is employed for the calculation of steady current in an electromagnetic field; a requirement is a constant time variable, as well as a charge that is a subject of neither a build-up or depletion

   The formula:

   **representations: B = electromagnetic field outside a circuit

   μo = permeability of o space

   I = electric current in an enclosed circuit

   R = distance from the electromagnetic field

   $B = \int (\mu o \times I) \div 4\pi R^2$

3. Faraday's Law – like his [Michael Faraday] law of induction, this is a law that addresses the induced electromagnetic force of an object; it is strictly applicable to the charged elements within a closed circuit

   The formula:

   **representations: IEMF = induced electromagnetic field

   Dφ = difference of space

   DT = time differential

   $IEMF = - D\varphi \div DT$

4. Lorentz Force – a law that assesses a point charge due to both an electromagnetic field and an electric field

   The formula:

**representations: LF = Lorentz Force

q = total charge

EMF = electromagnetic field inside a circuit

V = velocity

B = electromagnetic field outside a circuit

LF = q [EMF + (V X B)]

VII.D. – Electrodynamics & Quantum Electrodynamics

Before the 1900s, a scientist named William Gilbert addressed a proposal concerning *electricity* and *magnetism*; according to him, while both subjects can be traced to attracting and repulsing objects within a circuit, electricity and magnetism are different concepts. The key to understanding electromagnetism, therefore, is to understand the individual terms; particularly, understand their relevance and distinction.

A conflict regarding electromagnetism is that, despite agreeing to Albert Einstein's *Special Relativity*, it goes against some of the rules of mechanics; it is only dependent on the electromagnetic permeability of 0 space. It follows that in the case of moving frames, the electromagnetic field is subjected to transformation to include space.

Moreover, all electromagnetic phenomena are covered under *quantum mechanics*. This makes the electromagnetic field of a circuit accountable for the physical phenomena that are observable, especially *magnetism* and *electricity*.

Chapter VIII – Let's Talk Circuit Boards

Are you familiar with *crocodile clips*?

Crocodile clips are devices that can be used for the assembly of a circuit; these tools are so-named for their resemblance to the jaws of a crocodile. With them, a solid grip to connect a component is possible. If you're wondering how electrical engineers can create electrical connections without the associated dangers? Well, there's your answer.

By using *crocodile clips*, you can make a model of a working circuit. Whether be it a basic circuit, a series circuit, or a parallel circuit, you can create for an audience to analyze.

VIII.A. - Printed Circuit Boards 101

Printed circuit boards (or PCBs) are devices that enable electrical connectivity even in an "open" environment; within their system, there are resistors, inductors, transformers, capacitors, conductors, and semi-conductors. These tools support high component density. In a way, printed circuit boards are referred to as *live circuits*.

Since functioning circuit boards can be rather risky, especially when exposed to extreme environments, they are packaged accordingly. In most cases, these devices are subjected to a series of coating procedures and are dipped in acrylic, wax, polyurethane, and epoxy.

Design standards:

- Templates and card dimensions are designed according to required circuitry regulations

- Manufactured Gerber data are generated

- Design is planned thoroughly with the assistance of an *EDA or Electronic Design Automation* tool

- Traces for signals are routed

- Copper thickness and layer thickness are carefully evaluated

VIII.B. – Circuit Board Tests

To see if a circuit works, certain tests are performed. Particularly, it is determined whether or not it is functional and can perform desired tasks. Along with its capacitors, resistors, transformers, and other components, it is analyzed for opens and shorts.

Objectives of testing methods:

- To detect flaws

- To detect error-free operations of each of its components

- To determine system stability

- To evaluate whether it is fit for use

- To evaluate safety issues

- To verify test systems

Example testing methods:

- Analog tests

- Contact tests

- Contact tests

- Electrolytic capacitor tests

- Flash tests

- Powered digital tests

- Short tests

VIII.C. – Let's Learn to Prototype

Prototyping is the ability to put a particular idea to test by preparing a model from which other circuits are developed. Especially when the circuit in subject involves a complex system or expensive components, a prototype is initially designed.

If the circuit creator is unsure of how a particular circuit will function, his best bet is to create a *prototype*. This way, he can evaluate his creation and see how it can be improved. If it's not yielding his desired results, he can modify the placements of each of its components until he achieves a necessary output. Otherwise, he can proceed to the actual circuit-making process.

VIII.D. - The Art of Bread-Boarding

Breadboards allow the creation of a circuit prototype, which is the reason why these tools are ideal for beginners. *Bread-boarding*, therefore, is the process of creating a circuit prototype in a board to resemble the operations of a real circuit. Its history can be traced to the time that

41

electrical enthusiasts would use a literal breadboard (i.e. the board used for slicing bread).

According to experts, it's recommended to learn and understand bread-boarding prior to making your first-ever circuit. It's advised to be familiar with each of its components, as well as how it works.

Breadboard components:

- Chips – these are *legs* that come out of both sides of a breadboards; these components fit perfectly and serve as connectors of different parts

- Posts – these are components that enable connections from power sources

- Power rails – these are vertical metal rows strips that are adjacent to terminals; through these components, easy access to a power source can be provided

- Power supplies – these are components that enable the supplementation of a wide range of electric current and voltage levels

- Terminals – these are horizontal metal row strips that are adjacent to power rails; through these components, wires are allowed to be inserted, then, be held intact

VIII.E. – Essential Skills in Circuitry

After understanding *bread-boarding*, you're almost ready for some hands-on circuit lessons. First, however, you should adopt a certain set of skills; while bread-boarding is helpful, it's only practice (i.e. for instance, there is no soldering involved). Since you'll be making an actual functioning circuit, it's time to get your hands dirty.

5 skills required in circuitry:

1. Stripping (wire) – it is a skill that promotes secure electrical connections; it involves the knowledge on various types of wires, thickness of wires, and how to maintain a solid grip; since exposure to electric current is a risk, it is best to check out the appropriate tools for wires

2. Drilling – it is a skill that focuses on the proper drilling (of holes) in a circuit, then, making sure that electrical connections can remain intact; it is a slow process that requires practice with accuracy and precision

3. How to test batteries – it is a skill that requires testing the capacity of a battery; particularly, it involves measuring the current load in open and short circuits

4. How to use a glue gun – it is a skill that takes advantage of a glue gun's conductive property; it teaches how to insulate and how to set a semi-permanent coating; especially when there is a need to strengthen the joints of a circuit, it comes in handy

5. How to use liquid electrical coating – it is a skill that focuses on the ability to apply liquid electrical coating where conventional electrical tape is likely to fail; it requires precaution since about 30% of the parts in the liquid coating are quite volatile

VIII.F. – The Secrets of a Solder

Soldering is a process that involves (at least) 2 metals (or any conductive material); the 2 metals are joined by flowing and melting. In the industry of circuitry, it is one of the most fundamental methods in circuit creation; it allows independent components to work as one.

A technique of good soldering revolves around the knowledge of the amount of heat that is applied. For *basic soldering, 361F* is the temperature to consider and for *advanced soldering*, the goal is to arrive at a temperature of somewhere between *361F to 419F*.

In *metallurgical engineering*, a term named *flux* is commonly used; in circuit engineering, as soon as the circuit creation process commences, it will be introduced, too. It is cleaning agent and a flowing material that facilitates the soldering process. Should there be invisible impurities (e.g. oil, dirt, etc.), they will be removed for the purpose of not risking the integrity of a circuit.

Moreover, it is important to note that in soldering, the proper application of flux is suggested. Improper methods can result to joint failure. The system's damage (due to incorrect flux application) may not be obvious at first, but gradually, it is capable of corrosion and rendering a circuit useless.

Chapter IX – Sufficient Safety

On a sunny day, try heading outside of your house, then, checking out the electrical connections (wires) from one post to another. More likely, one, two, or even a flock of birds are calmly resting on power lines.

Do you ever wonder why they don't get shocked?

No, birds are normal creatures; they do not possess extra-special powers. In their heads, they understand that it's a must not to step on open electrical networks, which is one of the things that need to be covered for beginners in circuit engineering.

IX.A. – The Lack of Electrical Safety Courses

While some *circuit engineering professors* teach enough lessons about *electrical safety*, others are quite behind on the area. They assume that practical knowledge, as well as a general "be careful" would suffice; since the beginners in circuit engineering are frequent exposure to electrical devices, no, the statement won't cut it. Therefore, it is important to stress out the need to be careful especially during their first hands-on activity.

Electrical safety tips:

- Treat each electrical device as if live current is running inside it (regardless if you're aware that there isn't)

- Overloading sockets for a circuit board? Not a good idea

- To cut off running current, add a residual current

- Always disconnect (and not just turn off) electrical devices when working on a component

- Always turn off any electrical device if not in use

- Regularly check the conditions of sockets and plugs

- Practice extreme caution when dealing with liquids and electric current

- Make sure your hands are dry

- Avoid octopus connections (i.e. devices that enable multiple plugs or sockets)

- Keep electrical wires and cords tucked neatly

- Wear the proper attire when creating a circuit; put on non-conductive gloves and footwear with insulated soles

IX.B. - The Importance of Hands-on Circuit Lessons

Hands-on learning is important for a beginner in circuits. It gives light to the fact that in circuit engineering, it is more about actual field work. Once the concepts are understood, it's best to move on to creating functional circuit boards.

A common problem that is encountered by beginners? Sweaty hands. Even in an air-conditioned laboratory, there are people who have to deal with sweaty hands when handling circuits. Alongside, they have to deal with the need of maintaining a strong and solid grip on various tools. Although this may be a concern, it's one that requires practice.

IX.C. – The 80-20 Rule of Circuit Safety

In circuitry, there is a rule that focuses on the installments of transmitters in hazardous areas for devices with a circuit; it is called the *80-20 Rule of Circuit Safety*. According to the rule, it is recommended to take extra precaution when dealing with particular portions of an electronic device; there are some parts that increase the risk of electrical shock when *touched* or *moved*.

Hazardous areas of a circuit (that require extra precaution):

- Point with high impedance

- Point with high resistance

- Area near the voltage source

- Area near an opening

- Area with conductive elements

IX.D. – Troubleshooting Concerns

In the event that a device with a circuit is not function correctly, it is recommended to have it evaluated accordingly. While a beginner may handle the task of checking, a professional in circuitry is the one who is advised to assess the system. Since an expert is already familiar with different circuit components, and he knows exactly where to look for possible faults, he is more qualified for the job.

Troubleshooting tips:

- Determine whether the connections are secure

- Determine whether the wires are correctly connected to one another

- Check for circuit components that seem out of place

- Check for circuit components that may be larger in size

Chapter X – Here's a Multimeter for You!

An important and indispensible tool in circuitry is a *multimeter*.

Imagine a situation when a circuit project was presented to you. Since you were requested to detail accurate measurements for its electric current, voltage, and resistance, you go to the nearest equipment laboratory to borrow a stack of devices for assistance.

Now, imagine the same situation, but this time, you have this tool; instead of having to go to the equipment laboratory to borrow a stack of devices to measure the electric current, voltage, and resistance, you turn to that tool. This particular time, you have a multimeter.

If you're not familiar with the operations of a multimeter, let it be your first job prior to creating a circuit project. It's actually quite easy to use. So long as you are attentive to instructions, and you know its components, certain restriction, and the technique to maximizing its functions, you're set.

X.A. - What's a Multimeter?

A multimeter, also called a multitester, *Volt-Ohm millimeter*, and *Volt-Ohm meter*, is an electronic device that measures the electric current in a circuit; it is capable of measuring voltage, resistance, and a variety of other units in a circuit's system. To know how to use the device, it's important to be familiar with each of its components; if you can understand how the components work, you can also understand the operations of the entire device.

Parts of a multimeter:

1. The selection knob – it enables a user to choose a particular setting that is subject for measurement

2. The display meter – it enables the display of the measured reading; it can contain up to 4 digits, as well as a negative sign

3. The probes – these are plugged into a multimeter that can interpret and convert measurements from a device into a multimeter; usually, these come in a pair of red and black probes

   Types of probes:

   I.   IC hook

   II.   Alligator clips

   III.   Test probes

X.B. - A Multimeter in the Works

Inarguably, a multimeter is one of the most useful tools in circuitry and one of the tools that can make the task of building circuits easy. It can be used to measure *any object or device* that contains a circuit and an electric current. Using it, however, comes the hard part; in fact, in the world of circuits and electronics, it earned the title as one of the most challenging jobs. It may be effortless to get a reading, but getting an accurate reading, then interpreting the reading is another story.

How to use a multimeter to measure (a basic example):

1. Prepare an AAA battery.

2. Plug the black probe of a multimeter to the negative side (i.e. the side with a "-") of the AAA battery.

3. Plug the red probe of a multimeter to the positive side of the AAA battery.

4. Check the display meter; as recommended, check the meter twice.

5. List down the reading and begin the interpretation.

X.C. – Resolution, Accuracy & Input Impedance

The smallest part of a multimeter's scale is called as the *resolution*. It is responsible for achieving an accurate reading and interpretation. In many multimeter kinds, especially digital ones, it can be configured or *calibrated*. And, as the rules go, a device with low resolution doesn't require much completion time; a device with high resolution can require a demanding processing time.

Meanwhile, the *accuracy* of a multimeter refers to an error in the measurement of an electric current, in comparison to a perfect reading. It is relative to the device's resolution since resolution may not be calibrated accordingly if the absolute accuracy level is questionable. Therefore, to determine the total accuracy of a multimeter, its relative accuracy should be added to its absolute accuracy.

Formula for the computation of total accuracy:

**representations: TA = total accuracy

RA = relative accuracy

AA = absolute accuracy

$TA = RA + AA$

When talking about a multimeter's resolution and accuracy, *input impedance* is a set of terms that needs to be acknowledged, too. This is due to the device's inability to achieve accurate readings when it is not set accordingly. Especially for the measurement of a circuit's voltage, its input impedance has to be calibrated high (i.e. higher than a circuit's voltage) so the operation remains smoothly.

X.D. – Safety Concerns

Groups that are in charge of the manufacture of multimeters, as well as the authorities that promote the safe use of such devices have set safety standards. This is to emphasize the importance of the right employment of the electrical tools. Although the method for use can be rather straightforward, reckless habits can result to problems. Alongside the possibility of inaccurate readings, it can cause harm to the individual that is handling the devices.

Categories of safety standards:

- Category 1 – applicable to the employment of a multimeter, circuit, or any electronic equipment with a distance near main connections

- Category 2 – applicable to the employment of a multimeter, circuit, or any electronic equipment with a distance somewhere near the first phase of main connections

- Category 3 –applicable to the employment of a multimeter, circuit, or any electronic equipment with a distance near permanently installed loads

- Category 4 – applicable to the employment of a multimeter, circuit, or any electronic equipment with a distance near faulty current levels that can be quite high

Chapter XI - DIY Circuits: Simple Projects

When you're into circuits, as well as *electronics and electrical engineering*, you need to step up your game when it comes to building items. Isn't the main reason for learning the branch of electrical engineering for the creation of circuits, then, putting them to good use?

In the event that your first circuit project wouldn't turn out as desired, try not to get discouraged easily, and instead, give matters another go. Not getting the results you wanted maybe a bit of a downer, but eventually, the odds will be in your favor; look at the setback as an opportunity for learning. If your heart's into circuits, you'll soon get the hang of how things are done.

XI.A. – Common Tools in Circuitry

For the creation of a circuit, you can make use of just about any tool you come across; if you find equipment that will make you accomplish a task easier, then, maybe you should put it in your arsenal. You *can* use just any tool, yes, but, doing so is not advised. It is best to choose the right set (i.e. a set of tools with a non-conductive handle) to not put your safety at risk.

Common tools:

- Crimper

- Cutters (e.g. cable cutter, electrical cutter, etc.)

- Extraction tool

- Glue gun

- Non-metallic tweezers

- Pliers

- Screwdriver

- Soldering gun

- Wire-wrapping tools

XI.B. - Beginner Circuit Projects

Building your first circuit can be a challenge; since you're still a beginner, you may end up making mistakes. Maybe your circuit won't end up as functional as desired, despite having followed your understanding of a series of procedures. In such a case, this is where you have to tweak your work; be diligent in figuring out where you went wrong. If you're uncertain of your actions' impact, don't worry too much. The important part is to begin; you can, then, figure out the rest along the way.

7 easy circuit projects (derived from http://www.instructables.com):

- Project # 1 – Static Electricity Analyzer

    The Static Electricity Analyzer can detect nearby static electricity; to indicate that static electricity is present, its LED component glows. Apart from a detector, it can be used to analyze the electricity in its surroundings. It is an extremely sensitive device since it can even detect nearby hand movement without touching the antenna.

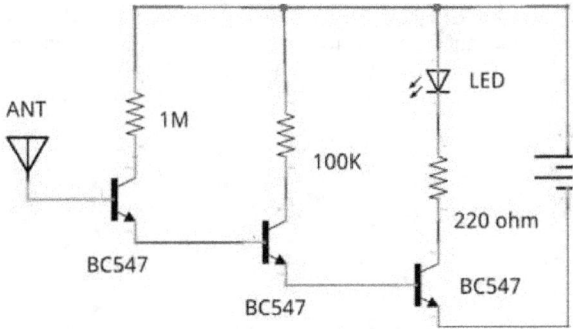

Materials:

- Around 10 pieces wire

- 1 piece LED

- 1 piece static electricity antenna

- 1 piece 100K resistor

- 1 piece 1M resistor

- 3 pieces 2n222 transistor

Procedures (as shown in the layout above):

1. To the left side of the circuit board, connect the 1M resistor to 2n222 transistor.

2. Beside it, attach the wire, then, attach the static electricity antenna.

3. On the bottom, attach the wire, then, attach the other 2n222 transistor; next to it, attach the wire, and finally, attach the third 2n222 transistor.

4. Place the 100K resistor adjacent to the second 2n222 transistor.

5. Place the LED adjacent to the third 2n222 resistor.

6. Establish connection to a power supply.

- Project # 2 – Dark LED Light

The Dark LED Light is a circuit project that can detect darkness. It follows that when insufficient light is supplied, an IC timer is alerted; consequently, a high output is produced and LED light will be switched on. The idea behind it is similar to that of a street light that automatically turns on once it detects that it's already evening.

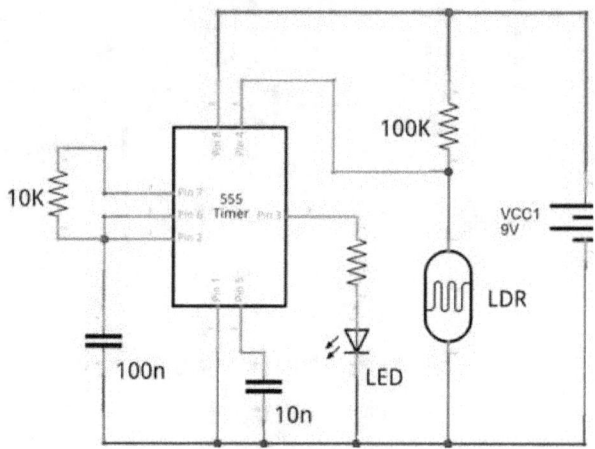

Materials:

- Around 10 pieces wire
- 1 piece LED
- 1 piece LDR

- 1 piece 10nf capacitor

- 1 piece 100nf capacitor

- 1 piece 10K resistor

- 1 piece 100K resistor

- 1 piece 555 IC timer

Procedures (as shown in the layout above):

1. On top of the LDR, attach the 100K resistor.

2. On the bottom of the LDR, create a connection to the LED.

3. To the left of the LED, create a connection to the 10nf capacitor.

4. To the left of the 10nf capacitor, create a connection to the 100nf capacitor.

5. Next to the 100nf capacitor, create a connection to the 10K resistor.

6. Place the 555 IC timer between all the connections; establish the main connection by attaching the other ends of the LED, LDR, capacitors, and resistors to the 555 IC timer (directly or with a wire).

7. Establish connection to a power supply.

- Project # 3 – The Ticking Bomb

The Ticking Bomb is meant for creating a ticking sound that resembles a bomb. Once it is turned on, it produces sound that is adjustable, but is modified to 1 tick per second.

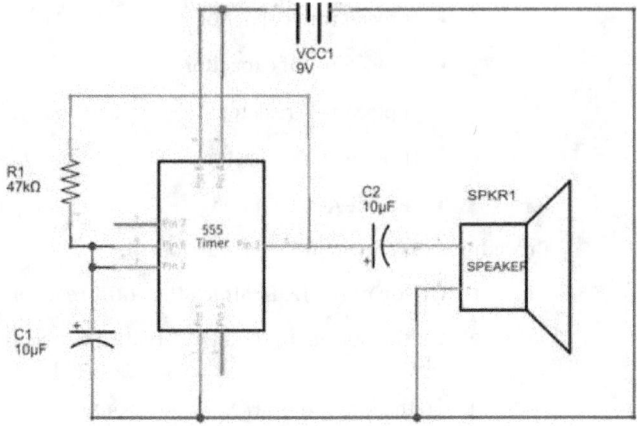

Materials:

- Around 10 pieces wire
- 2 pieces 10uf capacitors
- 1 piece 555 IC timer
- 1 piece 47K resistor
- 1 piece 8 ohm speaker

Procedures (as shown in the layout above):

1. Make the 555 IC timer as the central component; to its left, create a connection to the 47K resistor.

2. On the bottom, create a connection to the first 10uf capacitor; then, create a connection from the 10uf capacitor to the 47K resistor.

3. To the right of the 555 IC timer, create a connection to the second 10uf capacitor.

4. From the second 10uf capacitor, create a connection to the 8 ohm speaker.

5. Establish connection to a power supply.

- Project # 4 – The Remote Tester

The Remote Tester, as its name suggests, is a circuit that checks whether or not a remote control is working. Behind it, the idea is focused on the sufficient amount of signals that the IR receiver is getting. If it receives enough signals, the LED lights up, which means that a particular remote control is functioning; conversely, if the LED remains as is, it is an indication of a faulty component in the device.

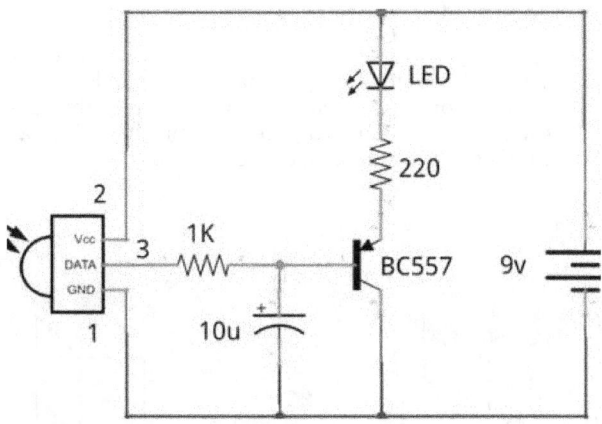

Materials:

- Around 10 pieces wire

- 1 piece LED

- 1 piece IR receiver

- 1 piece 1K resistor

- 1 piece bc557 transistor

- 1 piece 10uf capacitor

Procedures (as shown in the layout above):

1. On top of the bc557 transistor, create a connection to the LED.

2. Adjacent of the bc557 transistor (to the left), create a direct path to the 1K resistor, then, to the IR receiver.

3. On the bottom, create a connection to the 10uf capacitor.

4. Connect the 10uf capacitor (directly or with wire) to close the connection.

5. Establish connection to a power supply.

- Project # 5 – The Bell Experiment

The Bell Experiment is a basic project that produces a musical sound; the result is a device that may be similar to a doorbell. It works according to each of its components; if the resistor, transistor, and IC are triggered, the bunch sends a signal to the speaker, which will then, create the sound.

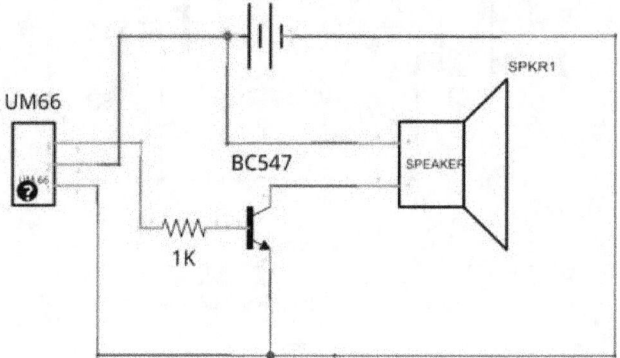

Materials:

- Around 10 pieces wire

- 1 piece 1K resistor

- 1 piece 2n222 transistor

- 1 piece UM66 IC

- 1 piece 8 ohm speaker

Procedures (as shown in the layout above):

1. On one end of the 8 ohm speaker, create a connection to the 2n222 transistor.

2. On another end, create a connection to the 1K resistor.

3. From the 1K resistor, create a connection to the UM66 IC; make sure that the UM66 is parallel to the 8 ohm speaker.

4. Connect the UM66 (directly or with wire) to the 2n222 transistor to close the circuit.

5. Establish connection to a power supply.

- Project # 6 – The LED That Fades

The LED That Fades is a project that produces and sometimes, blinking lights. It operates according to the weakness or strength of the signals that are interpreted by each of its components. If its IC timer, transistor, resistor, and capacitor receive strong signals; the LED will glow; conversely, if they receive weak signals, the light starts to fade. In the event that the signals that are submitted are unstable (i.e. they alternate between weak or strong in a few minutes' time), the blinking pattern comes in.

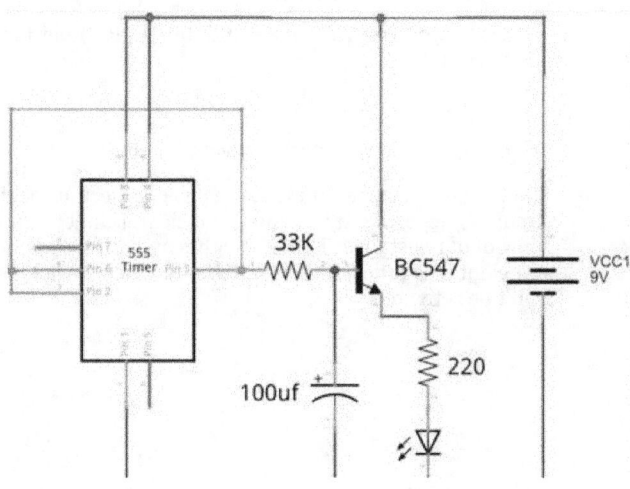

Circuit Engineering: The Beginner's Guide to Electronic Circuits, Semi-Conductors, Circuit Boards and Basic Electronics

Materials:

- Around 10 pieces wire
- 1 piece LED
- 1 piece 555 IC timer
- 1 piece bc547 transistor
- 1 piece 33K resistor
- 1 piece 220 ohm resistor
- 1 piece 100uf capacitor

Procedures (as shown in the layout above):

1. To the bottom of the bc547 transistor, create a connection to the LED.

2. Parallel to the LED, create a connection to the 100uf capacitor.

3. On top of the 100uf 33K resistor, create a connection to the 220 ohm resistor.

4. Next to the 220 ohm resistor, create a connection to the 555 IC timer.

5. Create a connection around the 555 IC timer for a closed circuit.

6. Establish connection to a power supply.

- Project # 7 – The LED with Activated Light

The LED with Activated Light is a basic project for beginners in circuit engineering; it is meant for a clearer understanding of the concept of resistance. The LED lights up with the application of sufficient resistance; if resistance levels are insufficient, the light won't be activated.

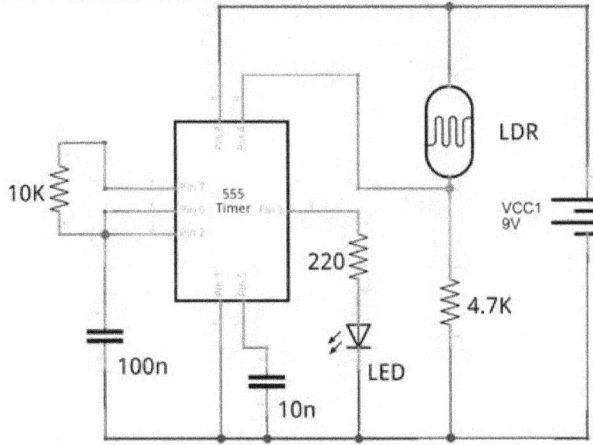

Materials:

- Around 10 pieces wire
- 1 piece LED
- 1 piece LDR
- 1 piece 10nf capacitor
- 1 piece 100nf capacitor
- 1 piece 10K resistor
- 1 piece 4.7K resistor
- 1 piece 555 IC timer
- 1 piece 220 ohm resistor

Procedures (as shown in the layout above):

1. To the bottom of the LDR, create a connection to the 4.7K resistor.

2. Parallel to the 4.7K resistor connection, create a new connection to the LED, followed by the 22 ohm resistor on top.

3. From the 22 ohm resistor, create a connection to the 555 IC timer.

4. To the bottom of the 555 IC timer, and parallel to the LED connection, create a new connection to the 10nf capacitor.

5. To the left of the 10nf capacitor, create a connection to the 100nf capacitor.

6. On top of the 100nf capacitor, create a connection to the 10K resistor.

7. Create a connection to close the circuit.

8. Establish connection to a power supply.

Chapter XII – Making Your Way to Circuit Design: PCB Layouts & Schematic Diagrams

If you are familiar with each of its components, you may have a good grasp of how an electronic circuit works; granted that the integral parts (e.g. power source, conductor, resistor, diode, etc.) are in place, and you were introduced to different techniques on establishing connections properly, there's a high chance that any circuit project you take on can be a success. Once an opening for a power source is identified, and the stability of an arrangement for the rest in a series is achieved, you're set.

However, if given the opportunity to make a particular circuit more functional, wouldn't you take it? Instead of settling for a random circuit design, why not have it modified for superior performance?

XII.A. - Circuit Design 101

Ever wonder why circuit engineers are paid highly and (depending on their chosen career path) are presented different career advancement opportunities (e.g. work as computer engineers, robotics specialists jobs, etc.)?

The responsibilities of circuit engineers are rather demanding; they need to be mentally tough and be open to different challenges. They spend hours and hours racking their brains out to determine how systems can be made more efficient. Chances are, there's always a way; it's up to them to look for one. When a particular project asks for it, it's mandatory for them to let their creative juices flow to arrive at a compatible solution. Otherwise, their built circuits may stop functioning eventually.

As mentioned, it's practical to design a circuit accordingly; especially with their employment for the fields of *communication, navigation, telecommunication, travel*, and other industries, their designs should consider a particular purpose. Apart from the objective of meeting different requirements, there needs to be a strategy since a more functional circuit comes with higher quality; it can make a project consume a reduced number of resources, too.

Reasons why there are various circuit designs:

- Improvement of a circuit's efficiency

- Improvement of a circuit's size and weight

- Guarantee a circuit's durability for a set period

- Guarantee a circuit is safe to use

XII.B. – The Design Process

Many times, a PCB layout (as discussed in chapter VII, *Let's Talk about Circuit Boards*) and a schematic diagram are often interchanged in circuitry; it has led others, especially the novices in circuit engineering, to believe that they are one and the same. In a way, since they are both presentations of a circuit's system, they may seem similar. However, upon closer inspection, they are not. It is, therefore, essential in the circuit design process that the difference is acknowledged.

A *PCB layout* is a physical representation or a *real model* of a circuit. It presents all of the electrical components that are included; it also details which of the components are active and which ones are passive. Although it can show an actual working circuit, understanding the functions of each of its parts can be tedious.

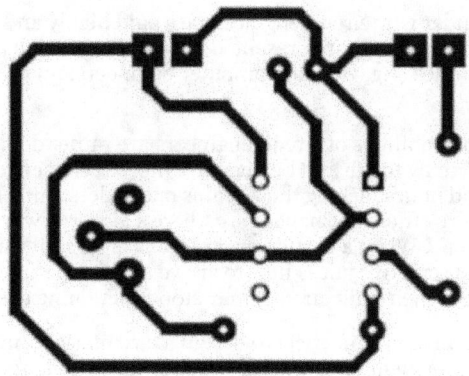

A PCB layout is an actual circuit model

On the other hand, a *schematic diagram*, also called simply as a circuit diagram, is a descriptive outline of a circuit's system; it is the standard and less costly way of circuit representation. It is like a PCB layout that shows both of the active and passive components in a circuit; unlike a PCB layout, however, its presentations can be easily understood.

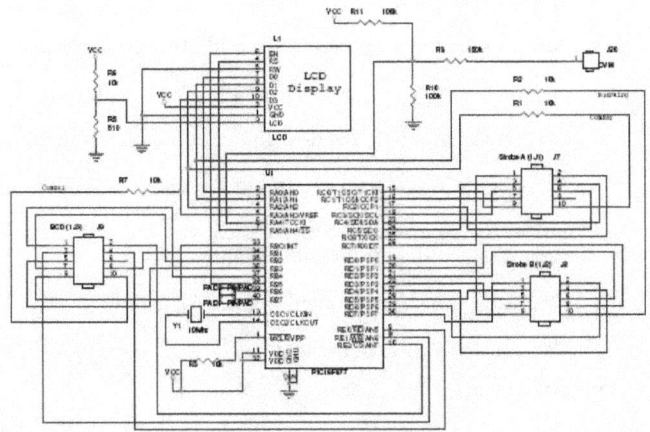

A schematic diagram can include important notes for the improvement of a circuit's design

Stages of circuit design:

- Meeting specific requirements

- The creation of a circuit diagram

- Building a breadboard, or a PCB layout

- The presentation of each circuit component (for professional evaluation)

- Applying the results of evaluation

- Testing (and retesting)

- Getting approval from professionals

Circuit design tips for beginners (preparation process):

- Categorize the components

- Construct a PCB layout, as well as a schematic diagram

- Determine the compatibility of each circuit component

Circuit design tips for beginners (circuit-building process):

- Always avoid cold solder joints

- Always separate power controls and other connections

- Always separate analog and digital components

- Create traces should hard-to-find components be included

- Remember to always make integral nodes accessible

- Solder the components systematically; solder small components first, then, solder larger components next

- Strategize the spaces you allow between components

- Take note of any heat spots

XII.C. - Circuit Symbols

On the process of building a circuit and modifying its design, an understanding of different circuit symbols is important. Descriptions of circuit components can be put in simple words; it is called a *verbal description*. However, since a thorough approach is recommended in many circuitry lessons, a visual representation can enable you to understand the electric flow in a circuit.

Here's a good comparison:

Verbal description	Visual representation
Circuit # 1 contains a light bulb, as well as D-cell battery as power source.	

Moreover, circuit symbols for a visual representation of a circuit is preferred over a verbal descriptions since they may be less complicated to understand. Especially if the particular circuit is quite advanced, a description containing words can be challenging to use as basis. If a visual model is employed, however, you can simply focus on connections, instead of the interpretation of a worded description; there is a better chance of building an impressive circuit design successfully. Especially if you have plans of advancing your place in circuit engineering by soon moving forward from a novice to a circuit expert, *memorizing* the different circuit symbols is a must.

List of basic symbols:

Symbol name	Symbol
AC	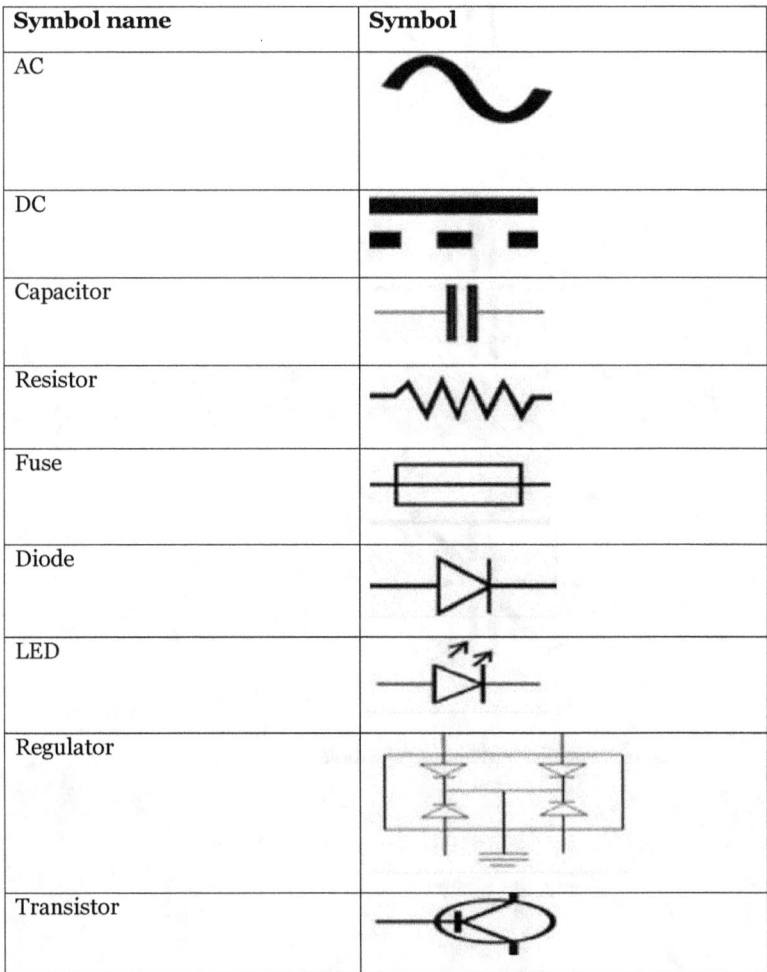
DC	
Capacitor	
Resistor	
Fuse	
Diode	
LED	
Regulator	
Transistor	

List of basic capacitor variation symbols:

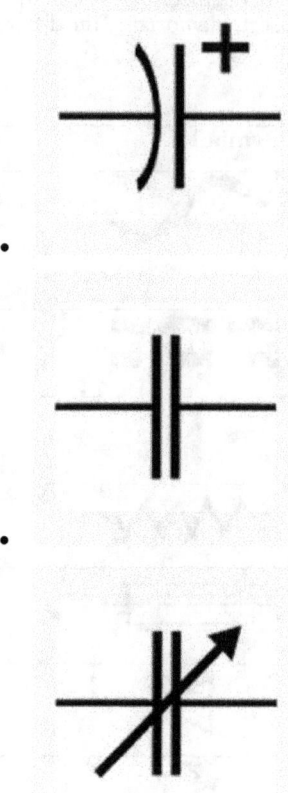

-

-

-

List of basic diode variation symbols:

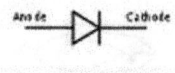

-

Ano de ▷◁ Cathode

-

Ano de ▷|| Cathode

-

Ano de ▷| Cathode

-

List of basic switches variation symbols:

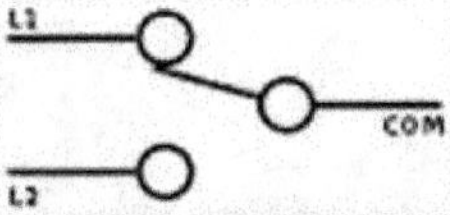

L1 ⟶ ○
○ ⟶ COM
L2 ○

-

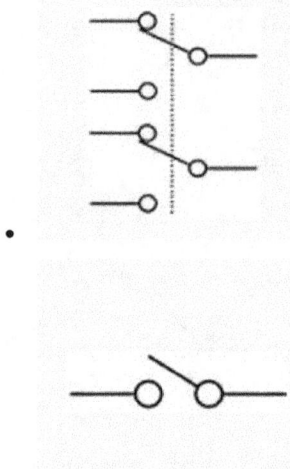

-

-

List of basic transistor variation symbols:

-

-

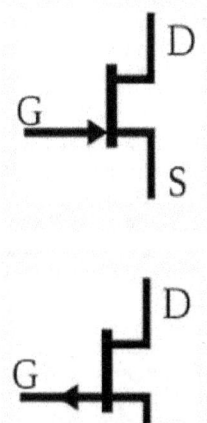

-

-

XII.D. – Factors to Consider

A strategic circuit design is important since a circuit may not function at all in the absence of a good plan; if the placement of each of its components wasn't accomplished according, you may have just wasted resources. You may have the correct components but if they are not designed properly, a circuit is far from working. Besides, at the realization that you may not be design a circuit's system according to requirements, you wouldn't want to put the circuit-building project to an end at the middle of the process, right?

Moreover, the proper design structure of a circuit needs to be prioritized. Regardless if it is not meant to impress another, it should still be designed accordingly as to be a stepping stone in the possibility of handling more projects eventually. With the knowledge of how to design a circuit properly as a beginner comes the effortless knowledge on the construction of more advanced circuits eventually.

What to keep in mind:

- Time availability (of the creator) – although it can be completed quickly, a circuit project shouldn't be rushed; the project creator should be available for a set period so he can concentrate properly on a circuit. Especially if he has hardly any time to spare, it's suggested that he either determines a way to find additional time, or takes on a different project

- Difficulty level – the assessment of a circuit project is important prior to acceptance. While there are simple ones, some projects are rather complicated. If a creator is certain that he can't follow through the procedures of a particular project, he may want to consider other undertakings

- Total cost of the project – one reason that the creation of a schematic diagram and other preparations are suggested (prior to taking on a circuit project) is due to practicality regarding the overall expenses. As much as possible, the creator should consider his ability to fund a project. Especially if he plans on using expensive circuit components, he needs to think about whether he can afford the project until completion.

- The desired function – identify the reason why you took on the project. Is it for experimentation and personal use? Or, was it a request from a company? Especially if the purpose is for a commercial company, design a circuit to achieve first-rate quality.

XII.E. - Documentation, Verification & Testing

A primary benefit of the *documentation* process is its advantage of letting you identify possible errors in putting together a circuit. If you weren't successful at the end stage of your project, you can figure out exactly where you went wrong; in favor of having to disregard your current progress and starting from scratch the second time around, you can simply have certain methods modified (for improvement).

Verification, as defined in circuitry, is the process of thoroughly evaluating each of a circuit's components, as well as each of the stages in the making. The objective is to determine whether the process has been adhered to correctly. Usually, this can be a time-consuming process, but since it is intend to ensure quality, it is an important one.

In a way, the real-world version of the verification process is *testing*. Prior to a circuit's launch and its exposure to different commercial industries, it is tested in research laboratories. Like verification, it can also be a time-consuming process, and since it is the last process before a circuit is passed on to another source, it can be a labor-intensive series.

Chapter XIII – A Way Is through EDA

Are you familiar with *FABS* or *Semi-conductor fabrication facilities?*

FABS are places where circuit designs, even complex ones, can be put together; they are usually extensive since they are meant to accommodate bulk circuit projects. Inside them are various equipment for *EDA*, which is why circuit-production is rapid.

With the productivity level of FABS, semi-conductors are continuously becoming popular. And, since semi-conductors are components in a circuit, circuits grew in popularity, too.

XIII.A. - What Is EDA?

EDA, also known as *Electronic Design Automation* is a class of software equipment especially for designing a circuit or different electronic devices. It is categorized along with other modern tools used in circuitry; it eliminates the need to design a circuit by hand, which minimizes simple and critical errors, repair glitches, and suggest possible improvements.

Moreover, it's due to EDA that circuit creation and circuit production became faster. If you're concerned regarding the quality since the process is done in an automated fashion, be confident that each design underwent thorough analysis. Remember, the software equipment's design was specifically meant to address a circuit's system.

XIII.B. – Design Flows

In the world of circuitry, it's important to be informed of the concepts of EDA; it privileges a circuit creator with a good grasp of the proper way to build circuits. The employees of grand electronics companies such as *Intel, Hewlett Packard*, and *Valid Logic Systems* are all trained to have impressive knowledge of the tools, despite already having gone through years of education.

Through the years, EDA tools have undergone continuous development. Initially, the concepts were rather limited; in the *Age of EDA Invention*, the focus was merely on *routing, logic synthesis, static analysis*, and *placement.* Eventually, in the *Age of EDA Implementation*, these concepts were subjected to major improvements; more *sophisticated and advanced algorithm* for the EDA tools were considered. Come the *Age of EDA Integration*, they were further studied and were designed to cater to integrated environments.

Now, you may be wondering why there is a need to build circuits manually when there are EDA tools. Well, although the process can be accomplished easier and faster, sometimes, bringing the expertise of a circuit engineer on the table is recommended; this way, he can thoroughly evaluate the quality of a circuit's system.

Additionally, the operators of EDA tools are selected carefully; usually, they are the ones with a background in the operations of EDA tools and similar equipment. This is due to the expenses, as well as the possible complications, involved in using such devices. In the event that a need for troubleshooting arises, it's best to have somebody who is knowledgeable in circuit designs on standby.

Primary focuses of EDA:

- The presentation of schematic-driven layout

- Advanced and logical interpretation of each circuit component

- Hardware and transistor simulation

XIII.C. - Circuit-Level Optimization

The circuit-level optimization or *power optimization* of EDA tools refers to the various techniques that are employed for the reduction of the total power in a circuit. Although the intent is to modify a circuit's setup economically, the processes shouldn't compromise the product's overall quality. With the main goal of enhancing a circuit's efficiency, the other objectives are to increase speed, eliminate possible leakage, enhance power distribution, and improve functionality.

Circuit-level optimization techniques:

- Modification of voltage scales, thresholds, variables, blocks, supplies, and other voltage-related concerns

- Modification and re-sizing of transistors

- Modification of logic styles

- Re-routing of networks

XIII.D. - Interpretation, Analysis & Verification

Although the automation process of EDA tools may eliminate errors in a circuit's system, a product remains a subject of various testing methods. This is due to the industry of circuitry's insistence on guaranteeing the flawless networks of circuit systems. In the aim of making sure that a circuit's functions are in tune with specific requirements, it is interpreted, analyzed, and verified.

In the event when a circuit (that was created with EDA tools) hadn't undergone tests, there is a likelihood that it won't function for a specific purpose. Especially if the project will be employed for commercial agenda, the industry experts are not granting permission to distribute "unevaluated" circuit systems.

Circuit evaluation methods:

- Assessment of a circuit's maintenance requirements
- Inspection of desired and undesired effects (in relation to mathematical logic)
- Inspection of a circuit's functionality
- Inspection of a circuit's stability
- Verification of physical components
- Verification of static timing

Chapter XIV – The Other Way around: Reverse Engineering

One of the coolest ways of understanding circuits is to disassemble all of its components one by one. First, make sure that it is not connected to any power source. After putting on protective gloves, begin taking apart a component; then, determine its function. Repeat the procedures until you know the purpose of the parts and the importance of having them work as a network.

Now, put them all back together; make sure the circuit works.

Such a process is called *reverse engineering.*

XIV.A. - An Introduction to Reverse Engineering

Reverse engineering, also called *deconstruction, reversing, backwards engineering,* and *back engineering* is the process of extracting knowledge from any device or electronic equipment. It involves the need to deconstruct an entire circuit for further analysis of its components. Although the methods that will be employed are opposite to the process of creation, it is considered as a practical approach of learning circuitry.

Moreover, reverse engineering gives light to Aristotle's concept that the key to understanding the operations of a device is studying each one of its parts. Back in the days, the field is limited to its employment on a circuit and other electronic equipment; now, in the modern world, the application of a "reverse" methodology includes almost anything – from children's toys and household appliances to neuroscience, computer programming, and DNA.

XIV.B. - Reasons for Reversing

One of the privileges that reverse engineering can grant to a circuit creator is the guarantee that a device is made accordingly. In the event that he chooses to base his circuit project on another circuit that was created with *questionable privacy*, he can determine any unethical practices. If he takes apart all of its components, he has the chance to have an insider look; by then, he can confirm whether unregulated modifications are in place.

Why use reverse engineering techniques for circuit-creation:

- Documentation purposes – especially in the case of shortcomings and low-quality circuit documentation, the diagnosis of a circuit is necessary to present new information

- Interfacing – with reverse engineering techniques can be subjected to interfacing; it can be evaluated accordingly, regarding its compatibility with another circuit

- Bug fixing – if there are critical faults in a circuit's design, it can be recognized better with a closer look at each of its components; instead of resorting to assumptions, a circuit creator can identify which part of his project needs modification

- Advanced technical information – advanced technical information is rewarded when opening up a circuit; especially for beginners, reverse engineering is a practical way of learning about the straightforward details within the system

- Incorporation of a new functionality – reverse engineering can allow the incorporation of a feature in a circuit; rather than design another circuit, a circuit creator can simply make modifications on his current project

- Modernization – reversing is beneficial in circuitry since it can be used to modernize a project; for instance, if most modern devices are popular in the market due to an innovative feature, a circuit creator can employ reverse engineering to add the same feature to his circuit

- Guarantee product security – if a circuit creator is unsure of the security of his project, reverse engineering is a way for him to determine certain concerns; he can check the specifications of each component and guarantee that none poses safety risks when put to use

XIV.C. – Is Reverse Engineering Similar to Hacking?

Since it is the method for the extraction of information on a circuit that can't be retrieved ordinarily, reverse engineering is argued to be a form of *corrupt hacking*; it is, therefore, a field that a few others in the electronics industry attempt to avoid. However, for many number of circuit engineers (as well as other engineers), there is brilliance in the entire concept of reverse engineering.

According to those who are not against the field, reverse engineering is a way of outsmarting an already finished product; additionally, as they would insist, isn't the point of engineering exactly that – to build and re-build until satisfying outcome is achieved? It may be considered as a form of hacking, but many contest to the idea that it is behind corrupt objectives.

XIV.D. – The Construction of Reverse Engineered Projects

Inarguably, reverse engineering is rather *destructive* and *invasive*. Some are not in favor of it since they do not welcome the idea of tearing their works apart. For others, however, it is a creative way of improving an already completed project; especially if the particular project is outstanding, they are granted the chance to make it even more outstanding.

Among the common projects that can be modified with the use of reverse engineering are *alarm clock radios, coffee machines,* and *colored lamps.* Alongside, the knowledge of internal systems, they can add a unique functionality that is not included during commercial distribution. For instance, you can add a new beeping sound to an alarm clock radio.

For the construction of various reverse engineering projects, it's advised to have a set of tools handy. Prepare a set of *screwdrivers, magnifiers,* and cutting equipment. Additionally, when the disassembly is completed, have a pack of *electrical tape* nearby.

Reverse engineering project tips for beginners:

- Take pictures of a circuit's front and back system prior to disassembly; you can use it as reference

- List down the set of procedures you plan on following; make sure you adhere to them

XIV.E. - Reverse Engineering & CAD

Reverse engineering rode along with the popularity of *CAD or Computer-Aided Design.* Through the years, the field that used to be limited to the basic improvement of a circuit's system began incorporating complex features. It provides a circuit creator the chance to analyze the internal portion of a device.

Moreover, instead of settling on getting a fundamental view of a circuit project that requires reverse engineering, a circuit creator can meticulously analyze a circuit; he can inspect each component thoroughly and come up at the most practical solution. For its development, circuit engineers, circuit designers, and other professionals on electronics started collaborative works with *architects* and those who are skilled in CAD.

Advantages of using CAD technology for a circuit project:

- To improve the alignment of each circuit component

- To enhance the designation of spaces within a circuit's system

- To modify geometric subjects for boosted performance

- To zoom in (and look closely) on each circuit component

XIV.F. - The Legality of the Industry

In light to the different discussions regarding its similarity to *corrupt hacking*, the reverse engineering industry is subjected to various legal complaints. There are even laws (that usually fall under contract laws and fraudulent manufacturing laws) meant for the protection of all sectors that employ *reversed* circuits or electronic devices.

The term *interoperability* is introduced in relation to a variety of legality concerns. With the emergence of cases that revolve around the disassembly of circuit's parts to compromise the quality of a circuit, some who tackle reverse engineered project are not received well.

Reverse engineering is, therefore, only considered illegal if the primary goal is to achieve interoperability. If the goal is for the improvement of a circuit's overall performance, it is encouraged; along with almost every other means of repairing a system, it is even recommended to arrive at a desired purpose.

Chapter XV – Hacking the System

Different communities of hackers host events that give light to those who are passionate in circuitry.

One community, *Artisan Asylum*, hosts a circuit hacking night once a week. In the gathering, circuit engineering fellows – from beginners to professionals, come together to discuss various concepts in circuitry and electronics. There, like-minded individuals share their love for circuitry and talk about their favorite projects, and basically, anything in the world of circuitry.

Moreover, Artisan Asylum's circuit hacking night provides great learning opportunities for circuit hacking enthusiasts. It presents lessons on how to solder, how to use particular computer programs, and how to modify a circuit to function as desired. Apart from teaching individuals the basics and advanced techniques on how to hack a circuit, and have it work as desired, the community encourages the attendees to think outside of the box and come up with brilliant and innovative ideas.

XV.A. – About Circuit Hacking

Circuit hacking, since it is linked with the word *hacking*, can sometimes be perceived as fraudulent. However, there is nothing fraudulent with it since the main reason why there are circuit hackers is for the improvement or the revision of a project; hacking, in this essence, is defined as the modification of an existing circuit to use it for a different purpose. And, in most cases, a circuit hacker is an individual who exhibits cleverness, open-mindedness, and technical aptness.

Due to a few similar concepts, circuit hacking and reverse engineering are said to be one and the same; they are not. While both may include certain techniques that are intended for the improvement of a circuit's operations, the former is merely focused on developing an existing circuit; it may be invasive, too, but it doesn't involve the *deconstruction of an entire circuit*. Especially if it was determined that the installment of a particular component can achieve a desired functionality, having to tear apart the other sections of a circuit is deemed unnecessary.

Common circuit hacking methods:

- Patching – a simple circuit hacking method that describes identifying a circuit's control mechanism or the most integral part of a circuit. Once the main component is identified, you can install a new and better component

- Component replacement – it is defined as replacing at least one component of a circuit with another component that comes with better quality

XV.B. – A Hacker's Main Tool: FIB Technology

FIB or *Focused Ion Beam* Technology is considered as one of a hacker's main tool since it grants him the chance to hack almost any circuit. Ever since the initial introduction of the applications in the 1990s, their usefulness hadn't come unappreciated. The early versions were not only quite expensive, but also, clearly, in need of improvement; later, the tools underwent continuous modifications from many electronics enthusiasts.

According to a study that was led by the engineers at Berlin Technical University, a person skilled in circuitry can install FIB Technology-based applications to hack into a system's security. For the particular research, an IC with low-level security was the focus; the objective was to work around its level of security with the goal of deliberately eliminating its defensive mechanism. The study was, of course, successful and eventually, it was proven that even high-level tools can be hacked with the same practice. And, as it follows, it sheds light on the concept that *there is no such thing as a tamper-proof circuit.*

Moreover, FIB Technology, as a clever technique of manufacturing, developing, and re-wiring a circuit, has earned the approval of different communities of hackers and circuit engineers. Alongside its advantage of boosting a system's performance, it reduces regular operation time. Due to its ingenious way of allowing an individual who's working on an electronic device to design (and even re-design repeatedly) his project, it was subjected to further developments.

Important parts of applications that incorporate FIB Technology (as shown in the layout):

- Aperture – it is in charge of gathering visual aids, then, modifying these tools for a clear display of retrievable information from a particular electronic device

- Deflector plates – it accepts, interprets, then, measures receivable data; initially, it acknowledges all information prior to the screening of the unnecessary ones

- Extractor – it is in charge of drawing out information from an electronic device, then, transferring them to the hacking mechanism

- Lens – it is in charge of making adjustments to assist when processing information

79

- Octupoles – it is also known as double quadrupoles or octopoles, which means something that has eight poles; it controls beams of ions

- Suppressor – it is designed to prevent power overload due to voltage spikes; it works by regulating the amount of electric current within a system so a device can remain functional

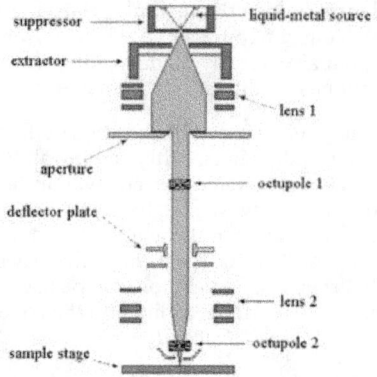

The installment of FIB Technology can be very promising to a hacker

Functions of applications that incorporate FIB Technology:

- Enable and/or disable intruder detection

- Evaluate a circuit's behavior

- Evaluate a circuit's defects

- Gather secret codes and security keys

- Obtain personal details, sensitive data, and proprietary information

- Remove protection systems (e.g. tamper networks, trace meshes, optical sensors, etc.)

- Route incoming data to be received by another network

- Trace and re-trace changes

XV.C. – Circuit Hacking Project

A plus side to the knowledge of how a circuit operates is that you can choose any from your pack of electronic devices and have it upgraded according to preference; the possibilities of its new functions are endless. You won't even have to spend a grand amount, so long as you're familiar with the functions of particular installments. The result of hacking the original system may be rather bizarre, since the product is no longer the same, but nonetheless, the result is likely the way you desired.

In the sample project below, the goal is hack a charger; particularly, modify an existing charger and have it operate with a battery. It can be connected to any electronic device with a USB port. This is useful during emergencies when an electric outlet is nowhere to be found.

Project example (derived from http://www.maximumpc.com): USB Charger with Battery

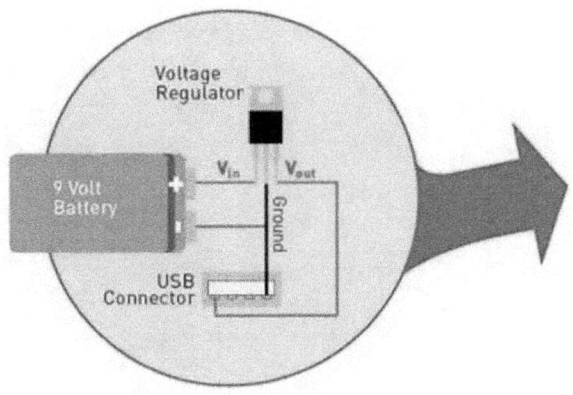

A USB charger with battery can be used for hours (depending on your voltage source)

Materials:

- Charger
- 5-volt voltage regulator
- 9-volt battery
- 9-volt battery clip
- Electrical tape

- USB connector

- Copper wire

Procedures (as shown in the layout above):

1. On a side of the charger, drill a hole for the placement of the USB connector.

2. On the other side of the charger, drill a hole for the placement of the 5-volt voltage regulator.

3. At the bottom center (between the 2 holes), place the 9-volt battery.

4. Above it, establish a ground using the copper wire and 9-volt battery clip.

5. Solder the components together.

6. Wrap the product with electrical tape.

Chapter XVI – Advanced Circuit Engineering: Microcontrollers & Robots

Among the plethora of prospects for a circuit engineer is the opportunity to engage in *circuit-bending* or the art (and science) of modifying existing circuits of electronic devices, and turning them into new musical instruments. In many cases, he isn't required to follow a set of rules for *tweaking* a circuit to incorporate a particular sound; in fact, he can re-design an electronic device as desired.

Alongside, an advantage of circuit-bending is its reward of reducing necessary expenses. In the event that he is very resourceful, the circuit creator can maximize the advantage even more; he can install used (but in working condition) components or less costly parts.

Circuit-bending is merely one of the exciting possibilities for a fellow in circuitry. The options are rather limitless, especially if you let your creativity run loose. So long as you are certain that a circuit will work given a particular arrangement, you shouldn't hold back in taking your beginner's knowledge of circuits to an advanced level.

XVI.A. – A Circuit Engineer's Future in Robotics & Computer Engineering

A reward of being skilled in circuitry is the opportunity to venture into other engineering fields such as *robotics engineering* and *computer engineering*. Your knowledge of how a circuit operates? You can look at it from a brand new perspective; you don't have to be simply in the industry of circuit engineering or electronics engineering. Apart from its offer of a more bountiful career; you can employ it to create a project (or a batch of projects) that you can be proud of. With the mastery of the basic circuitry lessons, delving into related fields becomes easier and more exciting.

With your interest in circuitry, you may seek for other career positions; usually, the employers of robotics engineers and computer engineers welcome circuit engineers into their workforce due to *trainability, familiarity with circuits,* and *good background in electronics.* It may take another set of years of studying, along with new skills to learn, but you can definitely go higher; it takes commitment from your end, too.

Advanced lessons that will be useful for a circuit engineer:

- Integral connections for hardware components

- Software and hardware essentials

- Robotic essentials

- Computer operations

- Computer architecture

- Computer programming (recommended programming languages are C and C++)

XVI.B. – Microcontroller + Microcontroller Programming

A *microcontroller* is a small device that serves as the computer in a circuit; it is a common tool that can be found in *remote controls, smart medical assistance equipment, office machines, state-of-the-art appliances,* and *engine control systems.* With the rapid pace of various information-retrieval operations, its function of addressing *size, cost, time,* and *performance* concerns offers privileges a user with improved overall performance. It is usually implanted on a separate electronic device before or after that device is finished. In certain cases, it uses 4-bit words and low frequency clock-rate operations.

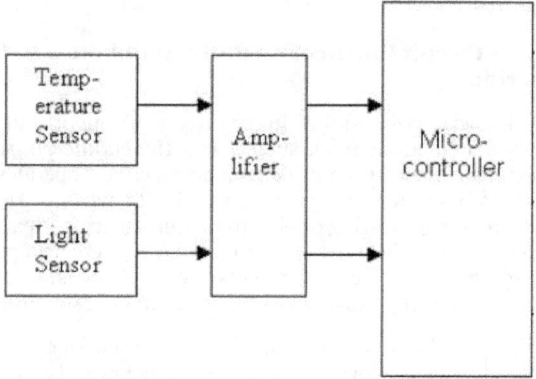

A microcontroller is dependent on a programmer's instructions

Moreover, while a microcontroller is a very useful device, it is only as good as the program that was written for it; this is where the importance of *microcontroller programming* enters the picture. It is a tool that merely executes certain instructions; in the absence of specific commands, it won't function. However, if you design it with even a hundred task capabilities, it can perform each one without fault; granted it was programmed well, it can power a device as desired.

What a microcontroller can do:

- Enable and/or disable clock feature
- Enable and/or disable light capabilities
- Configure audio and video settings
- Add an external monitor
- Automatically detect and/or repair errors
- Automatically update old components
- Magnify an electronic device's performance
- Eliminate unwanted features
- Tighten security features

Example of microcontroller programming:

Result	Microcontroller Programming (in C language)

By tweaking its USB components, a microcontroller's clock setting can be formatted. It is also programmed to activate BL or backlight with a trigger.	for (;;) [
	
	handleserial ();
	
	CDC_USBdevice (&interface)
	
	CDC_USBdevice (&digitalizer)
	
	handleserial ();
	
	]
	
	]
	
	/Setup/ [
	
	int ret
	
	USB_initialize
	
	clock_setup (div1);
	
	BL_off (1), BL_on (0);
	
	ret = digitizer_USB_initialize ();

	if (ret)

XVI.C. – Robots + Robotics

Another way that you can take your knowledge in circuitry up a notch is to consider a career in *robotics engineering*. Due to your familiarity with a circuit, you can begin honing your skills in creating coherent circuits; you can construct a series that is dependent on the individual circuits to deliver its primary function.

As you commit to robotics, like in programming, a lineup of new skills (e.g. artificial intelligence developments, dynamics, motion planning techniques, mapping tactics, etc.) must be acquired, too. This time, other than your hand in circuits and other electronic components, you have to visit aspects in the *mechanical* fields. However, given your exposure to similar concepts in circuitry, learning in the field may not be a challenge.

With your expertise in circuits, you can be behind innovative and extraordinary robotics projects that can be advantageous for both commercial and personal use. It just takes a matter of determination to move forward, and the fact remains that a course in circuit engineering can present you with a satisfying place in robotics, as well as other promising opportunities.

Conclusion

Thank you again for purchasing this book!

I hope this book was able to help you understand the fundamentals in circuit engineering. The lessons shared to you here are meant for a beginner in the subject; the different discussions are written simply. And, so far, it may have dawned on you that there's still more to discover about circuits.

The next step is to learn even more about circuitry and circuit engineering. Especially if you're considering a career in the field, advanced lessons would be good. Since this book has introduced you to the subject, and maybe inspired you to see the fun side in circuits, as well as electronics and electrical engineering, you may want to take the beginner's perspective to a whole other level.

Finally, if you enjoyed this book, please take the time to share your thoughts and post a review on Amazon. It'd be greatly appreciated!

Thank you and good luck!

Book 2
Cryptography
By Solis Tech

Cryptography Theory & Practice
Made Easy!

Cryptography: Cryptography Theory & Practice Made Easy!

Table of Contents

Introduction

I want to thank you and congratulate you for purchasing the book, *Cryptography*.

This book contains tips and techniques on how to build cryptosystems – even if you're just a complete beginner.

This eBook will help you learn about the history and basic principles of cryptography. It will teach you the different aspects of message encryption. In addition, you will learn how to establish cryptosystems. Aside from discussing modern/digital encryption schemes, this book will teach you how to use different types of "practical" ciphers.

Thanks again for purchasing this book. I hope you enjoy it!

Chapter 1: Cryptography – History and Basic Concepts

The Origin of Cryptography

During the ancient times, people needed to do two things: (1) to share information and (2) protect the information they are sharing. These things forced individuals to "encode" their messages. This "encoding" process protects the message in a way that only the intended recipient can understand the information. That means the data will remain secure even if unauthorized parties get access to it.

The art and science of protecting information is now known as "cryptography." The term "cryptography" was formed by fusing two Greek terms, "Krypto" (which means "hidden") and "graphene" (which means writing).

According to historians, cryptography and "normal" writing were born at the same time. As human civilizations progressed, people organized themselves into groups, clans, and kingdoms. These organizations led to the creation of concepts such as wars, powers, sovereignty, and politics. Obviously, these ideas involve information that cannot be shared with ordinary citizens. The group leaders needed to send and receive information through protected means. Thus, cryptography continued to evolve.

The Contributions of Egyptians and Romans

1. Egyptians

The oldest sample of cryptography can be found in Egypt. Ancient Egyptians used hieroglyphs (i.e. a system of writing that involves symbols and images) to share and record pieces of information. In general, these symbols and images are only intelligible to the priests who transmitted messages on behalf of the pharaohs. Here is a sample hieroglyph:

Fig. 1 - Egyptian Hieroglyphs

Several thousands of years later (around 600 to 500 BC), Egyptian scholars started to use simple substitution codes. This style of encoding involved the replacement and/or combination of two or more alphabets using a secret rule. This rule was considered as the "key" in retrieving the real message from the coded "garbage."

2. Romans

Ancient Romans used a system of cryptography known as the Caesar (or Shift) Cipher. This system depends on moving each letter of the message by a certain number (three is the most popular choice). To decode the information, the recipient simply needs to "move" the letters back using the same number. Here is an example:

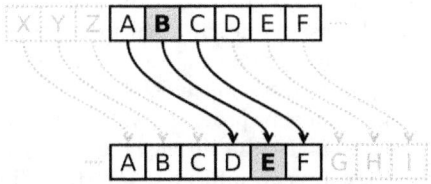

Fig. 2 – Caesar Cipher

Cryptography – Fundamental Concepts

Encryption – This is the process of converting information into an unintelligible form. It helps in securing the privacy of the message while it is being sent to the recipient.

Decryption – This process is the exact opposite of encryption. Here, the encoded message is returned to its natural form.

Important Note: Encryption and decryption requires specific rules in converting the data involved. These rules are known as the key. Sometimes, a single key is used to encrypt and decrypt information. However, there are certain scenarios where these two processes require different sets of keys.

Plaintext and Ciphertext – The term plaintext refers to data that can be read and used without the application of special techniques. Ciphertext, on the other hand, refers to data that cannot be read easily: the recipient needs to use certain decryption processes to get the "real" message.

Authentication – This is probably one of the most important aspects of cryptography. Basically, authentication serves as a proof that the information was sent by the party claimed in the encoded message.]

Let's illustrate this concept using a simple example: John sent a message to Jane. However, before replying, Jane wants to make sure that the message really came from John. This verification procedure can be conducted easily if John does "something" on the message that Jane knows only John can do (e.g. writing his signature, including a secret phrase, folding the letter in a certain way, etc.). Obviously, successful decryption of the message will be useless if the information came from an unwanted source.

Integrity – Loss of integrity is one of the biggest problems faced by people who use cryptography. Basically, loss of integrity occurs whenever the message gets altered while it is being sent to the receiver. Unnecessary and/or unwanted modifications in a message may cause misunderstanding and other issues. Because of this, the message must be protected while it is being delivered. Modern cryptographers accomplish this through the use of a cryptographic hash (i.e. a hash function that is extremely difficult to invert, modify, or recreate).

Non-Repudiation – This concept focuses on ensuring that the sender cannot deny that he/she sent the encoded message. In the example given above, it is important to make sure that John cannot deny the fact that he was the one who sent the message. Modern cryptography prevents this "sender repudiation" using digital signatures.

Chapter 2: The Modern Cryptography

For some people, modern cryptography is the foundation of technology and communications security. This type of cryptography is based on mathematical concepts like the number theory, the probability theory, and the computational-complexity theory. To help you understand modern cryptography, here is a comparison of the "classical" and "modern" types of cryptography.

Classical Cryptography

1. It utilizes common characters (e.g. letters and numbers) directly.

2. It relies heavily on the "obscure is secure" principle. The encryption and decryption processes were considered as confidential information. Only the people involved in the communication can access such data.

3. It needs an entire cryptosystem (will be explained in the next chapter) to complete the "confidential" transfer of information.

Modern Cryptography

1. It relies on modern technology like binary data sequences.

2. The information is encoded through the use of mathematical algorithms accessible to the public. Here, security is achieved through a "secret key" which is used as the foundation for the algorithms. Two factors ensure that "outsiders" cannot access the information even if they have the correct algorithm:

   a. It is extremely difficult to compute such algorithms. Thus, it is hard to extract any information from them.

   b. The absence or presence of secret keys.

3. The encryption doesn't involve the entire cryptosystem. Only the interested parties are required to participate in encoding and decoding the message.

Chapter 3: Cryptosystem – The Basics

A cryptosystem is the application of cryptographic methods and their appropriate coding environment (also called "infrastructure"). This system is used to add security to technology and communications services. Some people refer to cryptosystems as "cipher systems."

To illustrate the concept of cryptosystems, let's consider the following example:

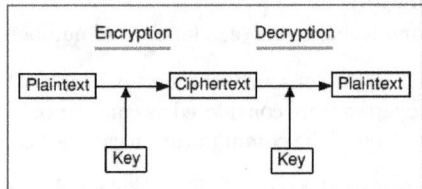

Fig. 3 – A Cryptosystem

In this example, the sender encrypts his message (i.e. the plaintext) through the use of a key. This process converts the plaintext into ciphertext. Once the receiver gets the ciphertext, he uses the same key to decrypt the information. Thus, the ciphertext will be turned into plaintext again.

As you can see, only the parties who have the "key" can access the shared information. Cryptosystems can be divided into these basic components:

- Plaintext – The information that needs to be shared and protected.

- Encryption Algorithm – The mathematical process that uses a plaintext and an encryption key to produce a ciphertext.

- Ciphertext – The coded version of the information. In general, cryptography protects the message itself, not the transmission. That means the coded message is sent through public channels such as emails. Thus, anyone who can access the selected delivery channel can intercept or compromise the message.

- Decryption Algorithm – It is a mathematical process that creates plaintext from any given set of ciphertext and decryption key. Basically, it reverses the encryption process done earlier in the transmission of the message.

- Encryption Key – The value provided by the sender. In creating the ciphertext, the sender enters this key (along with the plaintext) into the encryption algorithm.

- Decryption Key – The value known by the recipient. The decryption key is always related to the encryption key used for the message. However, these keys don't have to be identical. The recipient enters the decryption key and the ciphertext into the decryption algorithm in order to get the plaintext. The collection of all decryption keys for a cryptosystem is known as a "key space."

There are times when an interceptor (also called "attacker") tries to get the encoded message. In general, interceptors are unauthorized entities who want to access the ciphertext and determine the information it contains.

The Two Kinds of Cryptosystems

Currently, cryptosystems are divided into two kinds: (1) Symmetric Key Encryption and (2) Asymmetric Key Encryption. This way of classifying cryptosystems is based on the method of encryption/decryption used for the entire system.

The major difference between the symmetric and asymmetric encryptions is the connection between the encryption and decryption keys. Generally speaking, all cryptosystems involve keys that are closely related. It is impossible to create a decryption key that is totally unrelated to the code's encryption key. Let's discuss each kind of cryptosystems in more detail:

Symmetric Key Encryption

Cryptosystems that belong to this kind have a single key. This key is used to encrypt and decrypt the information being sent. The study of symmetric encryption and systems is known as "symmetric cryptography." Some people refer to symmetric cryptosystems as "secret key" systems. The most popular methods of symmetric key encryption are: IDEA, BLOWFISH, DES (Digital Encryption Standard), and 3DES (Triple-DES).

During the 1960s, 100% of the cryptosystems utilized symmetric encryption. This method of encrypting and decrypting information is so reliable and efficient that it is still being used even today. Businesses that specialize on Communications and Information Technology consider symmetric encryption as the best option available. Since this kind of encryption has distinct advantages over the asymmetric one, it will still be used in the future.

Here are the main characteristics of symmetrically encrypted cryptosystems:

- Before transmitting the message, the sender and the receiver must determine the key that will be used.

- The key must be changed regularly to avoid any intrusion into the cryptosystem.

- A stable form of data transmission must be established to facilitate easy sharing of the key between the involved parties. Since the keys must be changed on a regular basis, this mechanism may prove to be expensive and complicated.

- In an organization composed of "x" individuals, to facilitate two-way communication between any two members, the required number of keys for the entire system is derived using the formula: "x * (x-1)/2."

- The keys used are often small (i.e. measured through the number of bits involved), so the encryption and decryption processes are faster and simpler compared to those used for asymmetric systems.

- These cryptosystems do not require high processing capabilities from computer systems. Since the keys used are small and simple, ordinary computers can be used to establish and manage the cryptosystem.

Here are the two problems usually encountered when using this kind of cryptosystem:

- Key Determination – Before any message can be transmitted, the sender and the receiver must determine a specific symmetric key. That means a secure and consistent way of creating keys must be established.

- Trust Issues – Because all the people involved use the same key, symmetric key cryptography requires the sender to trust the receiver, and vice versa. For instance, if one of them shares the key with an unauthorized party, the security of the entire cryptosystem will be ruined.

Modern day communicators say that these two concerns are extremely challenging. Nowadays, people are required to exchange valuable data with non-trusted and non-familiar parties (e.g. seller and buyer relationships). Because of these problems, cryptographers had to develop a new encryption scheme: the asymmetric key encryption.

Asymmetric Key Encryption

These cryptosystems use different keys for encrypting and decrypting the message. Although the keys involved are dissimilar, they still have a logical and/or mathematical relationship. It is impossible to extract the message using a decryption key that is totally unrelated to the encryption key.

According to cryptographers, this mode of encryption was developed during the 20th century. It was developed in order to overcome the challenges related to symmetric key cryptosystems. The main characteristics of this encryption scheme are:

- Each member of the cryptosystem should have two different keys – a public key and a private key. When one of these keys is used for encryption, the other one must be used for decryption.

- The private key is considered as confidential information. Each member must protect the private key at all times. The public key, on the other hand, can be shared with anyone. Thus, public keys can be placed in a public repository. As such, some people refer to this scheme as "public key encryption."

- Although the private and public keys are mathematically related, it is practically impossible to determine a key using its "partner."

- When Member1 wants to send information to Member2, he needs to do three things:

    o Obtain Member2's public key from the public repository.

    o Encrypt the message.

    o Transmit the message to Member2. Member2 will acquire the original message using his private key.

- This mode of encryption involves larger and longer keys. That means its encryption and decryption processes are slower compared to those of symmetric encryption.

- The asymmetric key encryption requires high processing power from the computers used in the cryptosystem.

Symmetric key encryption is easy to comprehend. The asymmetric one, however, is quite difficult to understand.

You may be wondering as to how the encryption and decryption keys become related and yet prevent intruders from acquiring a key using its "partner." The answer to this question lies in mathematical principles. Today, cryptographers can create encryption keys based on these principles. Actually, the concept of asymmetric key cryptography is new: system intruders are not yet familiar with how this encryption works.

Here is the main problem associated with asymmetric key cryptosystems:

- Each member needs to trust the cryptosystem. He/she has to believe that the public key used for the transmission is the correct one. That person must convince himself that the keys in the public repository are safe from system intruders.

  To secure the cryptosystem, companies often use a PKI (Public Key Infrastructure) that involves a reputable third party organization. This "outside organization" manages and proves the authenticity of the keys used in the system. The third party company has to protect the public keys and provide the correct ones to authorized cryptosystem members.

Because of the pros and cons of both encryption methods, business organizations combine them to create safer and practical security systems. Most of these businesses are in the communications and information technology industries.

Kerckhoff's Principle
During the 19th century, a Dutch cryptographer named Auguste Kerckhoff identified the requirements of a reliable cryptosystem. He stated that a cryptosystem must be secure even if everything related to it – except the keys – are available to the public. In addition, Mr. Kerckhoff established six principles in designing new cryptosystems. These principles are:

1. The cryptosystem needs to be unbreakable (i.e. in a practical sense). This excludes the system's vulnerability to mathematical intrusions.

2. The system should be secure enough that members can still use it even during an attack from unauthorized entities. The cryptosystem needs to allow authorized members to do what they need to do.

3. The keys used in the system must be easy to change, memorize, and communicate.

4. The resulting ciphertexts must be transmissible by unsecure channels such as telegraph.

5. The documents and devices used in the encryption system must be portable and easy to operate.

6. Lastly, the system must be user-friendly. It should not require high IQ or advanced memorization skills.

Modern cryptographers refer to the second rule as the Kerckhoff principle. They apply it in almost all encryption algorithms (e.g. AES, DES, etc.). Experts in this field consider these algorithms to be completely secure. In addition, these experts believe that the security of the transmitted message relies exclusively on the protection given to the private encryption keys used.

Maintaining the confidentiality of the encryption and decryption algorithms may prove to be a difficult problem. Actually, you can only keep these algorithms secret if you will share them with a few individuals.

Today, cryptography must meet the needs of internet users. Since more and more people gain access to hacking information and advanced computers, keeping an algorithm secret is extremely difficult. That means you should always use the principles given by Kerckhoff in designing your own cryptosystems.

Chapter 4: Different Types of Attacks on Cryptosystems

Nowadays, almost every aspect of human life is affected by information. Thus, it is necessary to safeguard important data from the intrusions of unauthorized parties. These intrusions (also called attacks) are usually classified based on the things done by the intruder. Currently, attacks are classified into two types: passive attacks and active attacks.

Passive Attacks

Passive attacks are designed to establish unauthorized access to certain pieces of information. For instance, activities such as data interception and tapping on a communication channel are considered as passive attacks.

These activities are inherently passive: they do not attempt to modify the message or destroy the channel of communication. They simply want to "steal" (i.e. see) the information being transmitted. Compared to stealing physical items, stealing data allows the legitimate owner (i.e. the receiver) to possess the information after the attack. It is important to note that passive data attacks are more harmful than stealing of physical items, since data theft may be unnoticed by the receiver.

Active Attacks

Active attacks are meant to alter or eliminate the information being sent. Here are some examples:

- Unauthorized modification of the message.

- Triggering unauthorized transmission of data.

- Modification of the data used for authentication purposes (e.g. timestamp, sender's information, etc.).

- Unauthorized removal of information.

- Preventing authorized people from accessing the information. This is known as "denial of service."

Modern cryptography arms people with the tools and methods for preventing the attacks explained above.

The Assumptions of a Cryptosystem Attacker

This section of the book will discuss two important things about system attacks: (1) cryptosystem environments and (2) the attacks used by unauthorized parties to infiltrate cryptosystems.

The Cryptosystem Environment

Before discussing the types of data attacks, it is important to understand the environment of cryptosystems. The intruder's knowledge and assumptions about this factor greatly influence his choices of possible attacks.

In the field of cryptography, three assumptions are made about the attacker and the cryptosystem itself. These assumptions are:

1. Information about the Encryption Method – Cryptosystem developers base their projects on two kinds of algorithms:

    i.   Public Algorithms – These algorithms share information with the public.

    ii. Proprietary Algorithms – These algorithms keep the details of the cryptosystem within the organization. Only the users and designers can access information about the algorithm.

    When using a proprietary (or private) algorithm, cryptographers obtain security through obscurity. In general, these are developed by people within the organization and are not thoroughly checked for weaknesses. Thus, private algorithms may have loopholes that intruders can exploit.

    In addition, private algorithms limit the number of people that can join the system. You can't use them for modern communication. You should also remember Kerckoff's principle: "The encryption and decryption keys hold the security of the entire cryptosystem. The algorithms involved can be shared with the public."

    Thus, the first assumption is: "The attacker knows the encryption and decryption algorithms."

2. Obtainability of the Ciphertext – The ciphertext (i.e. the encrypted information) is transmitted through unsecured public channels. Because of this the second assumption is: "The attacker can access ciphertexts created by the cryptosystem."

3. Obtainability of the Ciphertext and the Plaintext – This assumption is more obscure than the previous one. In some situations, the attacker may obtain both the plaintext and the ciphertext. Here are some sample scenarios:

i. The attacker convinces the sender to encrypt certain pieces of information and gets the resulting ciphertext.

ii. The recipient may share the decrypted information with the attacker. The attacker obtains the corresponding ciphertext from the communication channel used.

iii. The attacker may create pairs of plaintexts and ciphertexts using the encryption key. Since the encryption key is in the public domain, potential attackers can access it easily. It's a "hit and miss" type of tactic.

Cryptographic Attacks

Obviously, every attacker wants to break into the cryptosystem and obtain the plaintext. To fulfill this objective, the attacker simply needs to identify the decryption key. Obtaining the algorithms is easy since the information is available publicly.

This means the attacker focuses on obtaining the secret decryption key. Once he/she gets this information, the cryptosystem is broken (or compromised).

Cryptographic attacks are divided into several categories. These are:

- BFA (Brute Force Attack) – Here, the intruder tries to find the decryption key by entering all possible information. For instance, the key contains 8 bits. That means the total number of possible keys is 256 (i.e. 2^8). The attacker tries all of these keys in order to obtain the plaintext. The longer the key, the longer the time needed for successful decryption.

- COA (Ciphertext Only Attack) – This tactic requires the complete set of ciphertexts used for a message. When COA gets the plaintext from the given ciphertexts, the tactic is considered successful. Attackers may also get the corresponding encryption key using this attack.

- CPA (Chosen Plaintext Attack) – This attack requires the attacker to work on the plaintext he/she selected for encryption. Simply put, the attacker has the plaintext-ciphertext combination. It means the task of decrypting the information is easy and simple. It is the first part of the attack – convincing the sender to encrypt certain pieces of information – that presents the most difficulties.

- KPA (Known Plaintext Attack) – With this tactic, the attacker should know some parts of the plaintext. He/she has to use this knowledge to obtain the rest of the message.

- Birthday Attack – This is a subtype of the brute force approach. Attackers use this tactic when working against cryptographic hash functions. Once the intruder finds two inputs that produce similar values, a collision is said to occur: the hash function is broken and the system is breached.

- MIM Attack (Man in the Middle Attack) – This attack is particularly designed for public key cryptosystems. In general, these systems require the exchange of keys before the actual transmission of the ciphertext. Here is an example:

  o Member1 wants to send a message to Member2. To do this, he sends a request for Member2's public key.

  o An intruder blocks the request and sends his own public key.

  o Thus, the unauthorized party acquires the information that will be sent by Member1.

  o To avoid detection, the intruder encrypts the data again and sends it to Member2.

  o The intruder uses his own public key. That means Member2 will see the attacker's key instead of Member1's.

- SCA (Side Channel Attack) – This attack is used to exploit the weaknesses of a cryptosystem's physical implementation. Here, the attackers ignore the system's algorithms and digital protection.

- Fault Analysis Attacks – When using this attack, the intruder looks for errors produced by the system. He/she uses the resulting information to breach the system's defenses.

- Timing Attacks – Here, attackers use the fact that different calculations require different processing times. These people can acquire some data about the message processed by a computer system. They do this by measuring the time used by the computer in performing its calculations.

- Power Analysis Attacks – These attacks are similar to the previous one. However, instead of time, they use the amount of power consumed by the computer system. This information is used to determine the nature of the plaintext.

An Important Note About Cryptographic Attacks

The attacks explained above are theoretical and highly academic. Actually, most of those attacks are defined by cryptography instructors. Some of the attacks you read about involve unrealistic assumptions about the attacker and/or the system's environment.

However, these attacks have excellent potential. Attackers may find ways to improve them. It would be great if you will still consider these attacks when designing your own cryptosystems.

Chapter 5: Traditional Cryptography

You have learned about the basics of modern cryptography. You also discovered the different tools that you can use in designing your own cryptosystems. One of the powerful tools at your disposal is the symmetric key encryption: a mode of encryption that uses a single key for the entire communication process.

This chapter will discuss this mode further so you will know how to apply it in developing cryptosystems.

Old Cryptographic Systems

At this point, you have to study the cryptosystems used in the ancient times. These "old systems" share similar characteristics, which are:

- These cryptosystems are based on the symmetric mode of encryption.

- The message is protected using a single tool: confidentiality.

- These systems use alphabets to facilitate encryption. In contrast, modern cryptosystems use digital data and binary numbers to encrypt a message.

These old systems are called "ciphers." Basically, a cipher is just a group of procedures performed in order to encrypt and decrypt data. You may think of these procedures as the "algorithms" of these ancient cryptosystems.

1. The Caesar Cipher

This cipher is based on a single alphabet. Here, you can create a ciphertext just by replacing every letter of the message with a different one. Cryptologists consider this cipher as the simplest scheme today.

The Caesar Cipher is also called "shift cipher," since each letter is shifted by a fixed number. If you are using the English alphabet, you can use the numbers from 0 to 25. The people involved must choose a certain "shift number" before encoding the plaintext. The number will serve as the encryption and decryption key for the entire communication process.

How to Use the Shift Cipher

1. The sender and the receiver select a shift number.

2. The sender writes down the alphabet twice (i.e. a-z followed by a-z).

3. That person gets the plaintext and finds the appropriate letters. However, he moves the letters based on the shift number selected. For example, if

they are using the number 1, he will replace the letter "A"s with "B"s, the "B"s with "C"s, and so on.

4. The encryption procedure is done once all of the letters have been shifted.

5. The sender transmits the ciphertext to the receiver.

6. The receiver moves the letters of the ciphertext backwards, depending on the shift number being used.

7. Once all of the letters have been shifted, the decryption process is completed. The receiver can use the information he received from the sender.

The Cipher's Security Value

This is not a secure system since the possible encryption keys are extremely limited. If you are using the English alphabet, your possible keys are restricted to 25. This number is not enough for those who need more security. In this situation, an attacker may acquire your key just by carrying out a thorough key search.

2. The Simple Substitution Cipher

This is an improved version of the Caesar Cipher. Instead of using numbers to determine the ciphertext, you will choose your own equivalent for each letter of the alphabet. For instance, "A.C... X.Z" and "Z.X... C.A" are two simple permutations of the letters in the English alphabet.

Since this alphabet has 26 letters, the total permutations can be derived through this formula: $4x10^{26}$. The people involved can select any of these permutations to create the ciphertext. The permutation scheme serves as the key for this cryptosystem.

How to Use this Cipher

1. Write down the letters from A to Z.

2. The involved parties choose a permutation for each letter. For example, they might replace the "A"s with "F"s, the "B" with "W", etc. These new letters don't need to have any logical or mathematical relationship with the letter they represent.

3. The sender encrypts the plaintext using the selected permutations.

4. The message is sent to the receiver.

5. The receiver decodes the ciphertext using the chosen permutations.

The Cipher's Security Value

This cipher is way much stronger than the Caesar Cipher. Even strong computer systems cannot decode the ciphertext since the possible permutations (i.e. 4×10^{26}) are too many. Cryptosystems based on this cipher can stop attackers that rely on a brute force approach. However, this substitution system is based on a simple scheme. In fact, attackers have succeeded in breaking letter permutations in the past.

The Monoalphabetic and Polyalphabetic Ciphers

Monoalphabetic Ciphers are ciphers that rely on a single encryption system. In other words, a single encryption alphabet is used for each "normal" alphabet throughout the entire communication process. For instance, if "C" is encoded as "X", "C" must be written as "X" each time it appears in the plaintext.

The two encryption systems discussed above belong to this type.

Polyalphabetic Ciphers, on the other hand, involve multiple encryption alphabets. The encryption alphabets may be switched at different segments of the encryption procedure. Here are two examples of polyalphabetic ciphers:

The Playfair Cipher

This encryption scheme uses pairs of letters to create encryption alphabets. Here, the people involved must create a table where letters are written down. The table used is a 5x5 square (i.e. 25 in total): the squares inside the table hold the letters of the alphabet. Since there are 26 letters in the English alphabet, a letter must be omitted. Cryptographers often omit the letter "J" when using this cipher.

The sender and the receiver must choose a certain keyword, say "lessons." They must write this keyword in the key table, from left to right. In addition, they should not repeat letters. Once the word is written down, the sender/receiver must complete the table using the unused letters (i.e. alphabetically arranged). With the word "lessons" as the keyword and J omitted, the key table should look like this:

L	E	S	O	N
A	B	C	D	F
G	H	I	K	M
P	Q	R	T	U
V	W	X	Y	Z

How to Use this Cipher

1. You should split the message into diagraphs (i.e. pairs of letters). If the total number of letters is an odd number, you should add a Z to the last letter. As an example, let's encrypt the word "human" using the key table created above. It will look like this:

HU MA NZ

2. Here are the encryption rules:

   a. If both letters are in the same column, you should use the letter under each one. You have to go back to the top if you are using the bottom letter. In our example, N and Z are in the same column. Thus, this pair becomes FN.

   b. If both letters are placed in the same row, use the letter located at the right of each one. You need to go back to the first letter of the row if you are working on the rightmost letter. (This rule doesn't apply to our example.)

   c. If none of the previous rules apply, create a rectangle using the pair of letters. Afterward, use the letters on the opposite corner of the correct letters. Work on the letters horizontally. According to this rule, the HU pair is converted to MQ (look at the key table). MA, on the other hand, becomes GF.

3. Using these rules, the word "human" becomes MQ GF FN when encrypted using the Playfair Cipher.

4. You just have to reverse the process if you want to decrypt the message.

The Playfair Cipher's Security Value

This scheme is stronger than the systems discussed earlier. Attackers will have a difficult time analyzing all of the possible keys. In general, cryptologists use this cipher to protect important information. Lots of people rely on the Playfair Cipher since it is easy to use and doesn't require special tools.

The Vigenere Cipher

This encryption scheme uses a word (also known as text string) as the key. This key is used to change the plaintext. For instance, let's use the word "human" as the key. You should convert each letter into its numeric value (i.e. A = 1, B = 2, etc.). In our example:

$$H = 8, U = 21, M = 13, A = 1, N = 14$$

How to Use this Cipher

1. If you want to encrypt "cold water," you have to write the letters down. Then, write the key numbers (i.e. 8, 21, 13, 1, and 14) under the words, one number for each letter. Repeat the numbers as necessary. It looks like this:

C	O	L	D	W	A	T	E	R
8	21	`13	1	14	8	21	13	1

2. Shift the letters of the normal alphabet according to the number written on the table. Here it is:

C	O	L	D	W	A	T	E	R
8	21	13	1	14	8	21	13	1
K	J	Y	E	K	I	O	R	S

3. As you can see, each letter of the plaintext is moved by a different amount – the amount is specified by the key. The letters of your key should be less than or equal to that of your message.

4. To decrypt the message, you just have to use the same key and shift the letters backward.

The Vigenere Cipher's Security Value

This cipher offers excellent security: better than the three ciphers discussed above. Cryptographers use this encryption system to protect military and political data. Because of its apparent invulnerability, security experts call this the "unbreakable cipher."

Chapter 6: Modern Cryptography Schemes

Nowadays, cryptographers use digital data to establish encryption systems. This data is often represented as sets of binary digits (also called "bits"). Modern cryptosystems must process these binary strings to create more strings. Symmetric encryption techniques are categorized based on the procedures performed on the digital information. These categories are:

Block Ciphers

These ciphers group the digital data into separate blocks and process them one at a time. The number of bits contained in a data block is predetermined and unchangeable. Two popular block ciphers, AES and DES, have block sizes of 128 and 64, respectively.

In general, a block cipher uses a set of plaintext data and produces a set of ciphertext data, usually of the same size. Once the block size is assigned, it can no longer be modified. The block size used for the system doesn't affect the strength of encryption techniques involved. The strength of this cipher relies on the length of its key.

Block Size

Although you can use any block size, there are some things you have to consider when working on this aspect of your cryptosystem. These are:

- Avoid small block sizes – Let's assume that a block size is equal to m. Here, the total number of possible plaintext combinations is 2^m. If an intruder acquires the plaintext data used for previous messages, he/she can   initiate a "dictionary attack" against your cryptosystem. Dictionary attacks are performed by creating a dictionary of ciphertext and plaintext pairs generated using an encryption key. You should remember this simple rule: the smaller the block size, the weaker the system is against dictionary attacks.

- Don't use extremely large block sizes – Large block sizes mean more processing time for your computer system. Cryptographers working on large bit sizes experience efficiency issues. Often, the plaintext must be padded in order to get the desired block size.

- Use a block size that is a multiple of 8 – Computers can easily handle binary digits that are multiples of 8. You can take advantage of this fact by choosing a block size that has this mathematical property.

Different Types of Block Cipher Schemes

Cryptographers use a variety of block cipher encryption schemes in their systems. Here are some of the most popular block ciphers being used today:

- AES (Advanced Encryption Standard) – This cipher is based on Rijndael, an award-winning encryption algorithm.

- IDEA – This is considered as one of the strongest ciphers available. Its block size is equal to 64 while its key size is equal to 128 bits. Many applications utilize this encryption. For instance, the old versions of PGP (Pretty Good Privacy) protocol used IDEA extensively. Because of patent issues, the utilization of this encryption scheme is restricted.

- DES (Digital Encryption Standard) – This is the most popular block cipher during the 1990s. Because of their small size, DES ciphers are now considered as "broken ciphers."

Stream Ciphers

With this scheme, the information is encrypted one binary digit at a time. The resulting ciphertext is equivalent to the data processed (e.g. 10 bits of plaintext produce 10 bits of ciphertext). Basically, stream ciphers are block ciphers with a block size limit of 1 bit.

Chapter 7: The Pros and Cons of Cryptography

This chapter will discuss the pros and cons of using cryptography.

The Pros

Today, security experts consider cryptography as one of their most useful tools. It provides four things needed for modern communication:

1. Authentication – Cryptographic techniques like digital signatures prevent spoofing and data forgeries.

2. Confidentiality – Encryption schemes protect information from unauthorized parties.

3. Non-repudiation – The digital signatures used in cryptography prove the identity of the sender. Thus, disputes regarding this factor are prevented.

4. Data Integrity – Cryptography involves hash functions that can maintain the integrity of the data being transmitted.

The Cons

Nothing is perfect. Here are the problems associated with cryptography:

*   Legitimate users may encounter issues when accessing well-encrypted information. This can turn into a disaster if a certain piece of information has to be obtained quickly.

*   Constant availability of information – a basic aspect of communications and information technology – is hard to secure when using cryptography.

*   Cryptosystems prevent selective access control. Some organizations need to provide exclusive file access to certain officers. When using cryptosystems, these organizations will have problems in giving selective access to their chosen members. This is because cryptosystems apply the same security measures on every file.

Conclusion

Thank you again for purchasing this book!

I hope this book was able to help you master the basics of cryptography.

The next step is to build your personal cryptosystems so you can easily encrypt messages.

Finally, if you enjoyed this book, please take the time to share your thoughts and post a review on Amazon. It'd be greatly appreciated!

Thank you and good luck!

Book 3
Open Source
By Solis Tech

Understanding Open Source From the Beginning!

Open Source: Understanding Open Source From the Beginning!

Table Of Contents

Introduction

I want to thank you and congratulate you for purchasing the book, *"Open Source: Understanding Open Source From the Beginning!"*

This book contains the basics in understanding the open source concept. What is it all about? Where did it come from? Who creates the open source content? How can software be considered as an 'open source'? What makes it different from the other software that we already have?

These questions are answered in this book. Also included in this book are information relevant to open source, such as examples of licensing, the Four Freedoms of free software use, and ideas about software piracy. This information will help to further understand what it means to have some software that is open sourced.

Real life comparisons are also made in this book in case you become confused or lost in understanding the open source concept. The idea of open source seems very simple, but in reality, it is very complex, with definitions coinciding with the definitions of other concepts such as free software (which will further be discussed in Chapter Two). Listed down in the book are the advantages and disadvantages of open source software, and the reasons why more and more people are becoming enticed with the idea of converting to open source.

If the present generation already dictates the movement of open source software, what will become of it in the future? This question is also answered in the last chapter of this book. Due to the fast-paced advancement of technology, open source will adapt to this advancement with the help of both developers and users.

Thanks again for purchasing this book, I hope you enjoy it!

Chapter 1: The Basics of Open Source

Have you ever wondered how an application you're using works? Every time you use an application and it freezes, do you think about what could have gone wrong? Do you ever think of why applications are constantly updating? These are questions that you would not be asked often. But these questions are very important to you, as a user of the Internet age.

Application programs are comprised of source codes, and these source codes are made by programmers. These codes are what allow you to type words into a word processing document, or to click on that video of cats meowing simultaneously. What you see onscreen are only visual representations of the codes of the program. Your application programs may be paid, or pre-installed in your devices, so you don't have permission to view these codes. Rather, you get the pre-made product, and you as a consumer have no power over it except to use it as instructed.

When you purchase or download an application and place it in your device, it installs a lot of files, but none of these files contain the source code. A software manager is included in your installed files to monitor the application as you use it. Whenever your application gets bugged or freezes, this software manager runs, and it prompts you to file a report to the software's developers to tell them exactly what happened. Once the report is filed, the developers study the bug, fix it, and release an update a few days or weeks later.

But what if you could see these codes for yourself? What if, whenever something goes wrong with the application, you could easily contact the developers or ask for help from other programmers easily? These questions are the foundations of open source, and you are about to learn more about it in the following chapters.

What is Open Source?

Open source is a computer program that has its source code visible to the public. The public – which we can refer to as the users – have the power to view, copy, and modify the source codes to their liking. The source code and the compiled version of the code are distributed freely to the users without fixed fees. Users of open source can pretty much do anything they like with the open source programs that they downloaded, since there are practically no restrictions.

To better understand the concept of open source software, let us use an example of recipes for comparison.

Recipes start off with someone writing them down on a piece of paper. A grandma, perhaps, has a recipe for a cake, which she writes in her recipe book. She passes on this recipe to her children, and tells them that they can use the recipe whenever they like. But, they must make sure to credit her as the original creator of the recipe.

The children recreate the recipe and whenever they are asked where the recipe is from, they would always tell that it's from grandma. One of grandma's children alters the cake recipe by adding strawberries as an extra ingredient. The grandma allows this, given that she is also permitted to use the altered recipe.

This example has the same concept with open source software.

When a programmer writes a code, compiles it into a program, and distributes both the source code and the compiled program to the users, he is giving everyone permission to access everything about the program. Users can run the program, view the code, modify if needed, compile, and redistribute the modified version of the program.

The original programmer, however, would require the users to let him use the modified versions of his program, since it is his to begin with. Aside from this certain restriction, the users of the program have the freedom to do whatever they like to do with it.

Let's go back to the example of the cake recipe. One of grandma's children, the one who added the strawberries, suggests to grandma to add the strawberries to the original recipe. The grandma thinks that this is a good idea therefore she complies and replaces her old recipe with the altered cake recipe.

In open source software, if the programmer is notified of a certain modification of a user, and it is deemed to be a modification that the software needs, then the programmer will revise his program based on that certain modification. This modification is called a patch. The user who has suggested of the modification is now coined as a contributor. This process of adapting user modification to an open source software is called upstreaming, because the modification goes back to the original code.

The concept of open source depends on the communication and collaboration between the software's developers and its users. Bug detection and fixing of open source is made easier because numerous users are working simultaneously to study the source code and to compile a modified, fixed version of the code.

With open source, it is not only the developers who are finding new ways on how the software can be improved and upgraded. The users can also contribute their ideas and knowledge in the upgrading of the software. The original developer or programmer can be called the maintainer who monitors the changes in his or her original software.

Let us then go back to the cake recipe. What if another child of grandma decides to do his own version of the cake recipe? He adds raisins to the cake recipe, and asks grandma if the raisins can be added with the strawberries in the original recipe. Grandma refuses, because she dislikes raisins. Instead of being disheartened, this child decides that he would create his own version of the recipe and share it with the people he knows.

If a certain modification makes no appeal to the developer, the one who suggested the modification may opt to make his own version of the program. This act of not patching a modification from the original program is called forking. A forked program is a certain program that alters the original program in such a way that it becomes its own program.

A forked program can be described as a chip off an old block, since it doesn't necessarily separate itself from the license of the program it originated from, although it may seem like it due to the avoidance of patching. Programmers that collaborate with open source result to forking if their modified versions of the original program are deemed unfit by the program's developer.

Nonprofit organizations are the prime developers of open source software. However, due to the freedom of customization that open source has given both users and developers, even large companies are adhering to the open source culture.

How did Open Source become popular?

During the early times of computing, software followed a protocol and design with everyone conforming to a certain cookie-cutter ideal. Software was yet to be imagined as cost-free, and the developers kept their codes to themselves. But then, during the early 90's, the idea of sharing one's code to the public became an accepted idea to most users. The concept of software being free and open sourced became a reality when, after decades, the likes of Mozilla Firefox and OpenOffice were created.

Open source rose in its ranks when developers started making open source alternatives of commercial software. These alternatives are free and can easily be downloaded from the internet, enticing most users to convert to open source. What made open source rise, however, was the idea of community. Fellow programmers could interact and communicate with each other, and even with the developers, which was unheard of during the early times of computing. People could collaborate with the developers of the software and share their insights.

Open source has also given its users the freedom to fully inspect software before they use it – an action that was impossible to do with closed source software. Users who are into coding try open source and study the code line by line.

The popularity of open source software has been anticipated due to the fact that a lot of people supported the cause. Programmers started creating open source projects to contribute to the cause, and users started to get accustomed to obtaining and downloading open source software. With volunteers signing up left and right, and organizations creating their own programs, the growth and expansion of open source software cannot be stopped anymore.

Chapter 2: History, Comparisons, and Relevance

Open source was not immediately implemented until the early 90's, where more and more people began to realize the importance of being able to share the source code of software without fees and royalties. Like any other idea, open source started out as a small thought of making software free for the public, and grew into the culture that it is today.

The History of Open Source: The Open Source Initiative

Eric Raymond, an American software developer, published an essay (turned book) entitled The Cathedral and the Bazaar in 1997. The essay speaks about two different types of software, which he labels the Cathedral and the Bazaar.

In the essay, Raymond describes the Cathedral to be the type of software in which with each release of software, the source code of the software will be available. However, with each build of the software, the certain code block that has been modified is restricted to only the developers of the software. The examples presented under the Cathedral type of software were GNU Emacs (a type of text editor) and the GNU Compiler Collection (a compiler that caters to different programming languages).

In contrast, the Bazaar is the type of software that has the Internet as the venue for their development, making the code visible to the public. The example presented under the Bazaar type of software was Linux (now a widely known computer operating system), in which Raymond coined the developer Linus Torvalds to be the creator of the Bazaar type of software.

Raymond's article became popular in 1998, getting the attention of major companies and fellow programmers. Netscape was influenced by this article, leading them to release the source codes of their internet suite called Netscape Communicator. The source code of the said internet suit was what gave birth to internet browsers such as Thunderbird and SeaMonkey. Mozilla Firefox, a popular web browser today, was also based from the source codes of Netscape Communicator.

The idea of source codes being free became widespread when Linux was developed, urging people to contribute to the open source cause. Because of the increasing popularity of Linux and similar projects, people who became interested in the cause formed the Open Source Initiative, a group whose advocacy is to tell people about the benefits of open sourcing and why it is needed in the computing world.

Open Source vs. Free Software

Most people confused open source software with free software, as the two terms share somewhat the same advocacy. With understanding, it is not that difficult to tell these two terms apart.

The difference between free software and open source software can be listed down into different points. Although they have their differences, both free software and open source software have a singular goal – to publicize source codes for the users to see.

Free software focuses mainly on the ethical aspect of the advocacy. There are certain freedoms that free software are fighting for when it comes to the use of software, which cannot be given to the users by commercial software. These are the Four Freedoms of software use according to advocates of free software:

•   The freedom to use the software. This means that the user is free to use the software to his or her needs, or as instructed.

•   The freedom to study the source codes of the software. Since the codes are readily available for public viewing, the user has the freedom to view and study the said codes. After he or she reviews the codes, he or she then has the freedom to do the next step.

•   The freedom to modify the source codes of the software to the user's liking. If necessary, the user has the freedom to customize the source code and to create a version of the program fit for the user's specific needs.

•   The freedom to share the modified, compiled source codes to the public. If the program has been modified, the user has the freedom to compile and publish the modified program for the benefit of the other users who may also have the need of the program's modification. The developer of the original program should also be given the freedom and right to use the modified version of the program.

Free software allows its users to do whatever they want with a program. If they want to modify the source code and redistribute the modified code as their own, without the consent of the original developers, then they are free to do so. If the user wishes to use the source code as the base code of a new project that they are working on, then they will not be sued. The ethical reasoning of free software simply states that there are no grave restrictions when it comes to copying, revising, and republishing the already existing software.

Open source, on the other hand, creates programs with the Four Freedoms in mind. The programs which are considered open source are made for the user's convenience and benefit. The common idea of open source is a group of people working on a single open source project, attempting to create a program that will be beneficial to them, as well as the users.

Open Source and Paid Software

Open source software did indeed come from paid software. There are countless of open source alternatives for common, commercially-sold software readily available on the internet. Some examples of this are office suites like LibreOffice and OpenOffice, which are open source alternatives for the much more popular Microsoft Office.

The reason why open source alternatives of paid software exist is mainly the cost. Users would opt to pay less, or none at all, for certain software. Why pay for software when there are free alternatives that can be downloaded from the internet easily? Open source makes it possible for users who cannot afford paid software to experience the basic and intermediate features of the software, without sacrificing the quality of the end product.

Although open source may be the overall solution for users to get a feel of certain software, there are still others who would want to obtain paid software but through illegal means. This is called software piracy, an action that is still evident despite it being illegal in most countries.

Software piracy is the act of downloading or installing a paid software illegally, either through software cracks or illegally burned CDs. The most popular way to obtain pirated software is through downloading Torrent-based software crack, in which the user can get the files through different computers almost discreetly. Since these software are pirated, installing these software requires the user to turn off his or her Internet connection before installing, to avoid being tracked.

Some paid software can be bought once, and shared with different computers or devices. All of the information regarding the sharing of paid software can be found on the software's End User License Agreement or EULA. The EULA is a splash screen shown at the start of the software's installation which contains the contract between the software's developers and the user.

The EULA may allow the user to share one copy of the software to different devices, or it may restrict the user from doing so. Once the user has violated this part of the EULA, it can then be considered as software piracy.

Something that a user should be aware of is a certain license called the GNU General Public License, the license that most open source software adhere to. The license permits the user to copy, modify, and redistribute the modified code, just as long as the source files and the original codes are still documented. This is important in understanding how and why open source software is allowed to move freely across the internet without being coined as software piracy, as compared to paid, propriety software.

With propriety or paid software, the user is buying only the license. He is not allowed to revise the code, to reverse-engineer the code, and to view the code by all means. The only thing that the user is allowed to do when he purchases propriety software is to use a copy of the software that the developer has provided. It may seem like an unfair deal to some people, because a user should be able to own something that he has paid for.

Open source software changes that idea. It gives the user the freedom to see the program's source code, letting the user know the program's 'skeletal system'. Even without paying for the software, the user gets the full potential functions of the software and not just an executable copy of it.

Importance of Open Source

Technology is rapidly changing. Experts are coming up with more ways to improve the lives of other people. It is the same with those who contribute to open source projects. Their advocacy is to create free programs that will benefit the users.

Open source is important in the evolution of quality software. With a lot of people contributing to one singular project, the software that is produced will be the best of its kind as it has been meticulously observed and reviewed by the contributors. Open source gives way for the collaborative effort of different programmers and users, with the users being secondary developers of a certain program. It is an interactive effort, with the users being able to update the program alongside the developers themselves.

The fast paced advancement of technology would often overwhelm content creators to the point that they would stop creating content altogether. Content creators who are left behind by technology's advancement are often working in small groups or on their own, and have no means of help from fellow creators of their kind.

With Open source, this is never the case. Each open source software has its own community to back a fellow programmer up during each build, ready to help out other programmers and users when needed. The open source community's bond with each other is what makes open source catch up with the fast advancement of technology.

Chapter 3: The Benefits and the Downsides of Open Source

The Benefits of Open Source

The most obvious perk of having open source software is the availability of the source code. With the source code available to the general public, people are able to study the code line by line. Students of programming can study the source code and implement some blocks of it into their own projects, honing their skills and improving their code. Users who are meticulous with their software can view the codes and customize the said codes to their liking.

Aside from the source codes being publicized, another perk of having open source software is that it is mostly free, depending on the software's license. Users of open source software do not have to pay a large sum of money to be able to enjoy the full functions of the software. If the license requires the user to pay, the user may still try out the software's full functions before purchasing.

Open source promotes community. If a user encounters a problem with the downloaded software, he or she can seek help from fellow programmers or the developers themselves through a forum. Users and programmers alike can communicate and share their experiences with using the software, helping other users to get used to the software. With other programmers keen on editing and revising the source code, updated and better versions of the software can easily be uploaded and shared within the community for the benefit of the other users.

Also, when something goes wrong with the open source software, the user has the option to fix the problem himself should seeking help be an option that is not convenient for him. In propriety software, this cannot be possible as the license and copyright prohibits its users from ever touching the program's source code.

If the user of propriety software does as much as reverse-engineer the product, then they could be violating the program's copyright and therefore, be taken to jail. Open source software removes this restriction from the users, giving them permission to fix solvable program problems on their own.

The benefits of open source software are not limited to personal use. Companies and businesses are adhering to the open source paradigm due to the endless possibilities at half the price or lesser.

More and more businesses are converting to open source mainly because it is more cost-efficient than purchasing commercial software. Companies also have more freedom with open source software in terms of customization, since they have the power to mold the software to fit their company's needs. These factors are beneficial in the growth and development of businesses in such a way that the businesses need not to put out a large sum of money just to be able to acquire a software that will be utilized in their business.

Open sourcing has become a way for people to have access to the things that they initially did not have access to. Users of software now have the ability to study the source code of the program they are using, and to know how exactly a certain function of the program runs by looking at its specific line of code.

A sense of community is also created between the software's developers and programmers from outside of their firms. Through open sourcing, the developers are able to communicate with other programmers with regards to how the software can be enhanced further.

Some users would say that using open source operating systems grants more security as compared to paid operating systems. For example, if a user installs the Linux operating system, he or she does not need an antivirus or a virus detection software to keep his or her files intact. The operating system itself has security measures for the user. This becomes a benefit for both professional and nonprofessional users because they have more room for important files rather than installing different kinds of applications for protection.

Open source software is made for the people, by the people. It hones itself to the needs and wants of each user. Because of this, there is no need for the user to upgrade his or her hardware every time the software upgrades.

Take Apple's OSX (operating system) for example. Certain updates of the operating system are available to download, with better features than the previous build. However, older versions of the Macbook and the iMac cannot avail of the recent builds as their hardware are not fit enough to accommodate either the size of the downloaded file or the features itself.

With open source software, the upgrades can be coded to fit each user's needs, depending on the user's hardware. If a certain upstreamed version of the open source software is available to download, different downloaders are made available by the developer with the specifications listed beside each downloader, catering to the different specifications of the user. The user himself can opt to customize the code of the program to be compatible with his device.

Allowing the user these freedoms over the software has given open source software a bit of a leverage over paid, propriety software. But then again, there will be nay-sayers who think that open source software isn't the way to go.

The Downsides and Disadvantages of Open Source

Open sourcing has given users lots of benefits, but it is not perfect. Some would still prefer paid software over any open sourced software. Here are some of the reasons why some users do not approve of open source.

Open Source: Understanding Open Source From the Beginning!

If a user is not in any way a technology expert, he or she would want software that is easy to use. Open source software is known to be more technical compared to their paid counterparts. Paid software focuses on its user interface, making the application easy for the user to understand the system. Open source software usually start out with a not so attractive user interface, but with the basic functions of the program intact. As the program gets updated with each build, the user interface changes and adapts to the needs of its users.

Most critics would say that paid or propriety software is still better in a number of factors as compared to open source software. Because more people are accustomed to using paid or propriety software, the idea that there are other types of software available is intimidating to them. People think that open source software is made only for the technology savvy users, with the interface hard for them to manipulate. Why download a complicated software when they can buy a simple, pre-made software that they are already familiar with?

Paid software has become a norm in the everyday lives of users. Large companies such as Microsoft and Apple have made their name known all throughout the world, creating technologies that users and consumers have grown to love. Because of their undying popularity, the rise of open source software is unknown to the general public. And even if they are known, those who are used to seeing the big names are hesitant to try out what open source might be.

Seeking technical help might seem simple with the numerous open source communities readily available, but it may sometimes be inconvenient to the user. Paid software offer professional tech support straight from the manufacturers.

Chapter 4: The Open Source Culture

Open source gives the user freedom to do whatever he or she wants in a software. Who wouldn't want the freedom to edit source codes to their own liking? With open source, this opportunity of customization is available at hand.

Why are more people converting to open source?

With the source code open for public scrutiny, looking for errors will be easier. Other software companies that do not have their source code publicized have their own set of programmers and developers figuring out the bugs in the software. This is an advantage for companies who always require their software to be updated regularly to keep up with the business.

Students who cannot afford the luxury of paid software turn to their open source counterparts to be able to utilize their functions without having to pay a large amount of money. Open source alternatives of Microsoft Office are available for the students to download should they need to use an office suite for their projects.

Some open source versions of paid software are actually better. Paid media players can play certain file types and extensions, but crash once the file extension is unrecognizable. Open source software developers take note of these bugs and create a media player that can play almost all media file types and extensions in high definition. Because of this, even users who are not actually technology savvy would convert to the open source alternatives of paid software just because they've heard and they know that they can get more out of the open source counterpart.

Programmers who want to practice their coding also rely on readily available open source software in their study. Because the codes of open source can easily be viewed and modified, programmers can base their project on open source software and publish it as their own, creating a program fork.

Businesses, on the other hand, turn to open source software for two main factors: cost efficiency and the power of customization. As mentioned in a previous chapter, with open source software readily available to download on the internet, the businesses do not need to spend a lot of money for a software that they cannot customize as their own. Open source gives them the opportunity to keep on upgrading their system as needed, therefore improving the quality of their software with each build.

The flexibility of open source software has enticed businesses to change to open source from propriety software. Businesses would often buy already existing software and attempt to use them as instructed by the developers. Open source software has its own rules and regulations, but if businesses want their software to be something specific, then the developers of open source software will deliver. With propriety software, the business is the one to adjust to the software that they have purchased, an action that is removed once businesses convert to open source.

Examples of Open Source Software

A wide variety of open source software are available for download. These software may be used for utility purposes, for multimedia purposes – anything that the user desires and requires. Here are a few examples of open source software that you as a user have probably heard of.

The prime example of open source software is an operating system called Linux. It is an operating system based off of UNIX that is available to different computer platforms and hardware.

Another example of open source software is the media player called VLC Media Player developed by the VideoLAN Organization. This media player can run a variety of multimedia files at high definition. Its paid counterpart is Microsoft's own Windows Media Player, which before its most recent build can only play a handful of file extensions.

When it comes to operating systems, Android is another popular example of open source software. A company called Android, Inc. (later bought by Google) has developed this mobile operating system using another open source kernel, Linux. It caters mostly to devices which have touchscreen on them, such as touchscreen desktop monitors, tablets, and smartphones, much like its counterpart from Apple called iOS. Android has its own application store called Google Play, where the users can install applications onto their phones mostly for free.

Netbeans, a well-known software developing application, is also an example of an open source software. It is a Java-created application that caters to different programming languages, and can be run on multiple operating systems. Programmers use Netbeans to create object oriented applications using the 24 programming languages that it caters to.

GIMP, or GNU Image Manipulation Program, is an Adobe Photoshop-like application that edits photos and creates graphic images. It has basic photo editing features such as cropping, grayscaling, and resizing, making it a simpler alternative to Photoshop. Like its paid counterpart, users of GIMP can also create animated GIF images, a feature that most multimedia artists are very fond of using.

Video and computer games can also be open sourced. Some open source games such as Tux Racer are available in the Linux package when downloaded. The principle of open source games is the same as any other open source software – the developers merging and collaborating with the users to create quality content to be distributed to the general public. However, the visual quality and elements of open source games are yet to be improved.

Other examples include PHP (a web development language), MySQL (used in databases alongside applications such as Microsoft Access and Microsoft Visual Basic), Python (programming language), Blender (an Autocad Maya-esque application that caters to 3D rendering), and many more.

Chapter 5: The Future of Open Source

What will happen in the future?

The future of open sourcing seems bright. With most businesses converting to open source software and most developers contributing to open source projects, the growth and expansion of open sourcing will continue. Open sourcing gives way for the innovation of modern software technology – with a lot of people working on one simple open source project, there is no doubt that the project will continue to be updated and improved.

Software will only continue to improve as time passes by. Open source software has made it easier for software to improve and upgrade itself due to countless of volunteers who are up to the challenge. While propriety software claim to start software trends, open source software advocates the upgrades of software that will be favorable to the needs of the users rather than to the bank accounts of the developers.

Open source software does not wish to waste the time and money of the user; rather, it aims to maximize both time and money, with the inclusion of effort, of the user when utilizing the software.

Presently, paid software are still dominant over open source software. Paid software have more leverage compared to open source software when it comes to reliability and familiarity, since they have been used by programmers and users alike for decades. There is still a certain percentage of users who are not aware that there are open source versions of their paid software, which they can help improve and customize to their own needs and liking.

More people will be aware of the benefits of open source software in the future. With propriety software releasing more licenses that restrict its users from certain software freedom, the existence of open source will lead to the users converting from propriety software due to the lack of free will.

In the future, there is a possibility that open source will be available not just for software, but also for other forms of content that have sources.

The future of open source as an idea or a paradigm will not be restricted to software alone. With the further advancement of technology, more and more gadgets will be locked down by licenses and warranties which restrict its users from fixing even simple problems that the product may have.

Gadgets are becoming more and more digitized, and copyright restricts people from ever touching or attempting to change the software. Because of this, some people are beginning to open up to the idea of open source not just for software, but also for hardware and gadgets that are used every day.

Open Source: Understanding Open Source From the Beginning!

Let us take tractors for example. Tractors are machines that are essential in farming. If a tractor breaks down, the farmer himself can fix the broken tractor and keep it running again without having to buy a new one. But the modernization of technology leads the manufacturers of tractors to add digital aspects into their products: tractors now have microchips and are operated via computers, therefore are now protected by copyright.

Now, if the new tractor breaks down, the farmer has no permission to fix the tractor himself. He must hire a specialist to fix the problem, or else he goes to jail.

Open source hardware has already started to rise in its ranks alongside open source software. It basically means that users are free to create their gadgets from scratch, using open source hardware. Although the idea seems taboo at present, the fact that gadgets are also being restricted from the users will give way for both open source hardware and software to rise even further, giving users complete freedom over the creation and implementation of the technology that they need.

Content creators are restricted from creating certain things just because of copyright laws. Even artists, who upload videos on websites like YouTube and Vimeo, get flagged just because of a certain song or a certain speech that had some sort of copyright over it. This restricts creative freedom. It also restricts the content creators from creating what they know and love, and sharing it with their viewers.

Will open sourcing become a culture in the future? Surely, with the massive amounts of information available for the users to share freely amongst themselves. Open source software has given way for an idea that will change the world of computing for everyone, and allows everyone to have access to the large chunk of information that was previously not available to them. Transparency when it comes to creating code and building machines will become a fad in the future, as more and more people are willing and able to create content and share it with other users.

Conclusion

Thank you again for purchasing this book!

I hope this book was able to help you to understand better the concept of open source and its benefits to the public.

Finally, if you enjoyed this book, please take the time to share your thoughts and post a review on Amazon. It'd be greatly appreciated!

Thank you and good luck!

www.ingramcontent.com/pod-product-compliance
Lightning Source LLC
Chambersburg PA
CBHW070254190526
45169CB00001B/404